AF697101

We become what we behold.
We shape our tools and then our tools shape us.

Marshall McLuhan

—

"For to set the mind on the flesh is death, but to set the mind on the Spirit is life and peace."

Romans 8:6

GOSPEL-CENTERED TECH

Raising Faithful Followers of Christ in a Digitally Distracted World

Nathan Sutherland

LION PRESS

Published by Lion Press
a division of Vilicus LLC
www.LionPress.org

Gospel-Centered Tech: Raising Faithful Followers of Christ in a Digitally Distracted World, First Edition

Printed in the United States of America

ISBN 978-1-962088-00-8 (paperback)
ISBN 978-1-962088-01-5 (e-book)
ISBN 978-1-962088-02-2 (audiobook)

For bulk purchases, please email cs@lionpress.org

Library of Congress Cataloging-in-Publication Data is forthcoming.

24 25 26 27 28 29 7 6 5 4 3 2 1

To Brad and Barbara

Contents

ACKNOWLEDGEMENTS

This book has been a wonderful reminder that God uses broken things to demonstrate his glory; writing, funnily enough, is not the same as speaking. It is through a labor of great love, and the contributions of many, that this book exists.

I am deeply grateful to the many supporters of Gospel Tech. From day one, you've listened to a podcast recorded from a kitchen table; you've given your time and treasure; you've prayed, encouraged, challenged, and celebrated this ministry (and therefore me). Your faithful support has been the grace of God to me, and allowed me to get this resource into the hands of parents and leaders. A special thank you to Blair Masenhimer, who cheerfully encouraged and equipped me to start writing.

It's not often that some of your best friends from junior high end up being life-long, deeply rooted friends, but I am lucky enough to count Ryan and Selena Frederick in this special camp of people. You both have blazed a trail in which I am grateful to follow, and I am thankful for the chance to create this resource. Additionally, without the edits and guidance of Anna Floit, this book would be illegible—thank you for bearing with me.

One of the greatest gifts of doing this work has been the Board of Directors I get to work with: John Bustad, Ben Jones, Ryan Frederick, Doug Rickabaugh, Erin Tangen, and my father, Greg Sutherland. Thank you for your faithful support, your steadfast prayers, and your refusal to let me get sidetracked.

I am indebted to the many teachers, leaders, and friends who have invested in me and in this project. Thank you for the conversations, deep questions, and patiently pointing me back to excellence (and sometimes, the proper use of verb tenses). Mrs. Aylard, I wish you were here to read this.

Some of my greatest inspiration comes from middle schoolers, and especially the ones I taught at AJH; you continue to inspire my teacher's heart with your humor, drive, and potential. It is because of many of the questions that you ask and the conversations we share that make me love this work.

And finally, my deepest thanks to my own humans, Owen, Henry, and Hadley, who regularly show me the passion God has given each of us, be it bugs, board games, or belly buttons; and to my wife, Anna, who bore with me through all the existential questions brought about through writing this book. All my love.

—March 2024

INTRODUCTION

Love God. Use Tech.

When I was a teacher, I was regularly asked by parents and guardians about what they should do with their child's technology. The subject came up at parent-teacher conferences, in meetings to set Individual Education Plans, in concerned emails, and often after report cards went home. Parents and guardians were concerned about smartphones, internet use, and gaming. They worried about sleep habits, grades, and shifts in attitude and appetite. Most of all, however, these caring adults worried that they were losing the bright and personable young person they once knew to the unknown dangers of the internet.

In these conversations with parents, my heart to see young people reach their potential in Christ met my passion for empowering parents to raise healthy youth in a tech world. In a few short years, our ministry Gospel Tech was born.

Gospel Tech exists to empower caring adults to talk about healthy tech, communicate the gospel, and connect the hope of the gospel to everyday tech lives. It's rooted in research, practice, and personal experience as a teacher, Christ follower, and parent of three littles growing up in this digital world.

These resources are simple: there are things to do, but ultimately this entire conversation rests on a single "been-done": In Christ we are new creations. That means this book can't be about simply making better tech users, because that isn't Jesus' goal. Jesus doesn't want You Version 2.0; he wants you to die to yourself and gain the life that makes you look more like him. That's the gospel (John 3:15–17; Ephesians 2:1–10; 2 Corinthians 5:17).

Gospel Tech empowers caring adults to fight for young people, not just with them, when it comes to tech. We can help them learn what it means to use tech from hope, not for hope, because they understand what it means to love God and use tech—and in that order.

What if we're talking about tech wrong?

Tech can be an overwhelming conversation. We're worried about our own kids and terrified of other people and their kids—which is understandable. The news and research on the subject doesn't help; it'd be comical if we weren't all so stressed trying to make sense of it. Here's a quick sampling of the top—contradictory—research on tech and children:

- There is no such thing as technology addiction, but tech addiction is growing and dangerous.
- Technology improves academics, motor skills, and behavior, but impairs our attention, reduces test scores, and lowers cognitive capacity.
- Gaming improves eyesight, but screens ruin our eyes.
- Tech improves attention span, but tech causes attention issues.
- We have evolved with technology to enhance our ability to think deeply, but modern tech inhibits our ability to think deeply.

- Screen time is no different from any other time, but screen time negatively impacts mental health.
- Bullying and predators aren't really a problem, but eleven percent of children experience unwanted online sexual solicitation.

No wonder we're confused.

As parents and concerned adults, we're left asking how to avoid tech's biggest pitfalls, how to game the system, and how to fix our children. I'm reminded of coaching junior high wrestling. Athletes would ask: "Coach, how do I get out of that pinning combo?" It's a solid question. Wrestling is built around imposing your will on an opponent. No one pins themselves—if someone gets pinned, it happens against their will. The answer for how to "get out" of a move is simple, if not easy: Don't get into it.

The same is true with tech. Many of our biggest tech problems come from companies successfully imposing their will on us. We may show up for fun, friends, or information but we often stay longer, pay more, and return more often than we planned. Companies are vilified for behavioral design, but it's also what has made them successful. It's time to be honest, and as social psychologist Jonathon Haidt writes in is seminal work The Anxious Generation, admit that "Unfettered access to everything, everywhere, at any age has been a disaster."

The problem isn't the screens, or even the time and money spent, necessarily. The problem is that much of it is designed to be unhealthy. This shift in understanding moves us from trying to fix tech (trying to get out of that pinning combo) and moves our attention on how to stay healthy in the first place. It's a shift from fear of tech toward hope for our children, rooted in the reality that God saves sinners—people like you and me—even in a world saturated with technology.

Why does the gospel matter for our tech use?

Much like our words, our tech use is an overflow of our hearts (Luke 6:45; Matthew 12:33–35). The gospel is the good news that God saves sinners, and this provides us with our starting point for all of our tech conversations because the gospel provides:

- A loving diagnosis
- A much-needed reality check
- And a simple response

The gospel brings with it the loving diagnosis of the bad news: We are sinners, dead in our sins, and without God's rich mercy, we can't change our status (Ephesians 2:1–3).

It's not news we can explain away, even if we try. And try we do. We say we're better than "those people"; we justify and explain away our own sin, while clearly seeing the moral failings of everyone around us. But it doesn't matter how good we are when our spirits are set on rebelling against God.

But here's the good news.

"But God, being rich in mercy, because of the great love with which he loved us, even when we were dead in our trespasses, made us alive together with Christ—by grace you have been saved" (Ephesians 2:4–5).

While we were still his enemies, God reached out to us. The gospel doesn't just forgive us our sins; it doesn't just give us eternal life; it changes our motivations by exposing our desperate need—and it offers our only solution. As the late Tim Keller said, "The gospel is this: We are more sinful and flawed in ourselves than we ever dared believe, yet at the very same time we are more loved and accepted in Jesus Christ than we ever dared hope."

Let's be clear. Following all the rules in the world—even using tech perfectly—has no ability to save us. In fact, the law brings wrath (Romans 4:15), and is the light exposing how deserving of judgment we really are (2 Corinthians 3:6). We need more than just good rules and solid enforcement of them for our kids. We need Jesus.

In Christ, we are forgiven and:

- Given new hearts (Ezekiel 36:26)
- Shown our purpose (Matthew 5)
- Made into new creations (2 Corinthians 5:17)
- Empowered by the Holy Spirit (John 14:16)

This is where the rubber hits the road in our conversations on tech. We are not saved from the world—to live apart from it, never having to deal with its broken people, misguided plans, or tragedies. We are saved for the world, called to be in it but not of it (John 17:14–16).

Only in the light of the gospel can our tech conversations become anything more than behavior modification. When we know and believe that we are sinners saved by grace alone, we can begin to love God and use tech. This shift in belief puts tech back in its "proper place," as Andy Crouch writes: "No matter what advertising says…the very best of life has almost nothing to do with the devices we buy. It has a lot to do with the choices we make, choices that our devices often make more difficult." Instead of tech serving as a distraction, escape, or digital palliative, it returns to its function: a tool to be used, enjoyed, and aid us in our work and creative endeavors. It's because of the gospel that we are then able to use our tech from a place of hope, not for our hope.

In this book, my goal is for parents (and other invested adults!) to be equipped in navigating tech conversations. We will begin with remembering the hope we have in Christ, we'll define terms so we can enter into meaningful conversations, and we'll walk through some tangible ways to connect the hope of the gospel to our tech use.

A Note on Endnotes

This book is an attempt to enter into a complex, on-going conversation on technology, parenting, and how our brains work—with the gospel anchoring it all. When I began this work nearly ten years ago, parents were resistant to entering this conversation, and reliable, ready research was difficult to find. This is no longer the case.

Much of my writing is inspired by the professional insights and vetted research of the brightest minds in the fields of neuroscience, psychology, sleep, and technology. And because I want this book to remain a conversation, accessible to all readers, I've chosen to use an endnote citation that eschews numbers in the text of the manuscript. All citations are given in endnotes, by chapter.

You will find a portion of the referred sentence followed by the full citation for context if you'd like to dive further into the content for yourself. My desire in choosing this format is to maximize the flow of information while also displaying the vast array of research and information that has been amassed over the last decade.

I'm hopeful that this citation formation, like the rest of the book, will encourage conversation, make reading more fluid, and push you to those endnotes—all with the greater goal of empowering families to take a gospel-centered approach to technology.

Part 1: A Gospel-Centered Approach

CHAPTER 1

God Doesn't Manage Sin

I grew up in a home that was part art exhibit, part circus. I am the middle of five siblings and the only boy. Between the dance productions and friendship bracelets I got my fair share of creative outlets, but not much say in what I wore for such events. If you were motivated enough, I'm certain pictures of young Nathan festooned with ribbons and tutus could be found. My high school graduation slideshow found one easily enough.

When I was eight, my dad brought home the original Nintendo. We played Mario Brothers and Duck Hunt, both on the same cartridge mind you, and something in my brain clicked. This is when my long, complicated journey with tech began. I couldn't put words to it for years but in that moment, I simply knew: video games were good. From that point on, until I was well into adulthood, I spent most of my free thought dedicated to getting back to that happy place—nestled in front of a screen, preferably with a sibling or two nearby, saving the world from yet another dragon-related incident.

My parents, sensing my growing obsession, wisely made

sure I always had something else to do. I went to church and school, made friends, was on the worship team, played a couple sports, and even joined the musical. Fun fact: Lederhosen aren't as comfy as they look. My body was busy being productive, but my heart was set on video games.

For the next twenty years I continued to game in relative moderation in terms of time. I rarely played video games more than a dozen hours a week, never lost a job, friend, or even failed a class due to gaming. Yet the day came in my late twenties when I realized I was going through the motions of life: daily work, teaching, and marriage were perfunctory as I waited for my next chance to game.

I daydreamed fictitious adventures with fictional people. I skimmed through the best things in life, the parts that promised to develop me into more of who God called me to be, and focused instead on what brought me instant yet fleeting satisfaction. Gaming became a sort of recreational medication.

When I had a bad day, I gamed.

When life got stressful, I gamed.

My go-to activity for a long weekend, holiday break, having friends over? Gaming.

When people asked that rhetorical question, "What would you do if you only had a week to live," the correct answer was, "Spend more time with family," but my heart said, "Eat junk food and play games."

The Reckoning

Then came the fateful day when my passion for unproductive escapism slammed into my reality that this life is not my own. It was 2:00 a.m. I was supposed to be asleep hours prior. I'd started shortly after 9:00 p.m. with the goal of a quick, satisfying completion of a long-awaited gaming task. "Quick," however, was foiled by that online terror of gamers everywhere: Other online players.

While my group that night might have a different version of events, I stand by my story that my team was terrible. They were so bad, I thought I'd just give it one more attempt. Then another. I played round after round, seeking satisfaction but getting more and more frustrated at the incompetence of my motley crew. Failing to see the common denominator between each of these losses and my rotating cast of teammates, my two-hour commitment stretched into four. I was growing frantic, yet no less determined.

The fifth hour was an abject failure. Not only did I not win, but I also now had four hours of shut eye before I had to be up for church. I knew there was a good chance my lovely bride would question why I was so haggard the next morning. With a heavy heart I climbed into bed and prayed a simple prayer: "Lord, I want to love you well. Please help my video games not get in the way of my marriage, my job, and the work you're asking me to do. Amen." I rolled over and quickly fell into a restless sleep.

And that was the last day I ever played a video game.

God doesn't manage sin.

I awoke from my fitful sleep four hours later, put on my happy face, and made it to church with my wife. The guilt and self-loathing had mostly washed away with the morning shower and commute and I was just beginning to feel like I'd pulled this whole thing off. Then the sermon started. I was only half-listening at first, what with the minimal sleep and dimmed lights, but something Pastor Steve said caught my attention. My addled brain tried to put it straight: *Was this a repeat sermon?* I was almost certain I'd heard it before.

Have you ever felt as though a sermon was planned just for you? You sit in the pew, running through the list of people who knew about your struggle, and trying to figure out who ratted you out? You feel so guilty, you start arguing with yourself about why you shouldn't feel guilty. I had all kinds of reasons:

"It's not that bad."

"I don't play that much."

"No one else is telling me it's a problem."

"I work hard—video games are my outlet."

For what it's worth, I've since found that the quality of my guilt is usually directly related to the quantity of my excuses. This particular sermon was on addiction, and my excuses were through the roof. It's worth noting that our pastor was an exegetical preacher—he spent three years teaching through the book of John. Yet this Sunday he jumped the tracks and tackled addiction.

In just a few minutes I had a proverbial barricade of excuses surrounding my head and heart, but God used a single line to break all my best defenses to bits: "If you've ever asked God to help you manage your sin, you're praying wrong. God doesn't manage sin. He kills it."

My words from the previous night rang in my ears, "Lord, please help my video games not get in the way..." God then graciously gave me a choice. It wasn't a new choice, but it was newly offered. He simply asked, "Do you trust me?" I need to emphasize here, God didn't say, "Quit video games." He asked me to reflect on whether I love him with all my "heart, soul, mind, and strength" (Luke 10:27). I knew that wasn't possible as long as video games had this grip on me. How could I love anything well when I was trying to get through a day to get back to gaming?

It's tempting to look at this story and say, "See, Nathan quit video games. My kid/husband/grandchild should too!" But giving up games, or any other tech, isn't the end-all. God doesn't just want our tech, he wants our hearts. If we don't submit our hearts to God we will simply continue to look for peace and purpose in created things rather than in the One we were made to enjoy. As Tim Keller said, "Our hearts are idol factories.

I gave up gaming and the rest has been unicorns and butterflies!

I do feel this story would flow better if I could say that everything after this was perfect. But if we skip forward a couple weeks, you'll find me standing in my backyard holding a shovel, contemplating why bad things happen to good people. As I dug my hole I stopped to whine at God. I had ten extra hours to fill in my week and I wasn't happy about it. The logical decision was to take it up with the Almighty.

"You mean to tell me this is the rest of my life? Wake up, go to work, come home, do more work, go to sleep, rinse and repeat?!"

God doesn't always respond to my childish outbursts, at least not directly, but His response to this tirade was very gentle, and may have come with just a hint of laughter. It was hard to hear over my complaining. "Do you trust me?" he asked again.

This wasn't about quitting video games. Unhealthy tech applies to any misuse of tech—not just games. It was about believing God was trustworthy. That he was better than games. We all know the right answer. But when it comes to trust, we can't just agree—we have to follow. My trust would only matter if I obeyed. Trust, in this case, meant walking away from games despite my list of reasons it didn't seem fair.

I asked God what to do with my extra time, and was quickly volun-told to lead my junior high's Young Life club (turns out ten hours a week is the perfect amount of time to run a club); this revealed that my passion for teaching extended beyond the classroom.

A few years later, I stepped out of teaching for the same reason I got into it: To see young people reach their full potential. And here I am now, writing these words in front of you today. I can say with complete confidence that if I'd decided to keep gaming eleven years ago, God would still love me, but I never would've written this book. I'd never have put in the years of work it took to start a non-profit, or learn to professionally speak, or do any of the arduous hard learning God would lead me to do.

Let's review this little tech testimony:

Video games are awesome.

It's science. But seriously, games are *designed* to be hard to put down. John Hopson, lead researcher for Bungie with the development of Halo, set a blueprint for what he termed *Behavioral Game Design.* Hopson notes the best games are highly repayable because they do three things:

1. Engage players using hope and rewards.
2. Create fear that keeps players from quitting.
3. Train players to come back to the game well after they have mastered the needed skill set.

When we wonder why games are so enticing, we don't have to look far beyond their designers. The teams of psychologists involved in game creation is what keeps us coming back for more. It's no accident that video games and casinos operate on the same principles: Their goal is to keep you engaged and take your time, focus, and money. Games might be awesome, but they are expendable if they distract us from God (Matthew 18:8–9).

God is way better.

Jesus tells a parable about the man who finds treasure in the middle of a field. The man's reaction is to sell everything he has and buy the field. This is a beautiful picture of us and God when it comes to even our best tech–our tech can be amazing, but God is way better. That man sold everything he had, but lost nothing. He sold a used donkey, some used clothes, an old hut and bought horses and oxen, regal garb, and mansions for his family with the treasure he found. He gave up everything, but lost nothing.

God calls us to do the same. We are to love him with all that we are—our hearts, souls, minds, and strength. It's not a temperate suggestion for moderate improvement; it's a divine prescription for joy and purpose. Holding anything back is tantamount to stealing from ourselves. Only in loving God wholly and completely can we truly know what it means to fully live.

We can trust him.

God isn't a cosmic kill-joy. He isn't just trying to teach us character by taking our nice things and making life harder. Instead he disciplines those he loves, and prunes the trees that bear good fruit (Hebrews 12:6; Proverbs 3:12; John 15:2). He is good. He is a good King, yes, but he's also a good Father. He understands what it is to suffer, to be fully human, and loves us even when we are actively running (Hebrews 4:15;

John 3:14–16; Romans 5:8; Ephesians 2:1–4; Colossians 3:1–4). Yet loving God isn't just a matter of will—it's a matter of grace. We can't force ourselves to follow God's laws, or to exude love any more than we can earn our way into God's good graces. And through his Holy Spirit, he promises to transform us to look more like Christ; as our hearts are changed, so too will our desires be changed.

What do we do with our unhealthy tech?

If you're struggling to find balance in your tech (or parenting a child who is struggling), know it's a tech issue *and* a heart issue. We'll talk more on this later, but video games and social media are designed to be unhealthy—so simply "playing less" may not be an option.

The great news is that this is exactly why Jesus came: He came that we might have life, and life to the fullest (John 10:10). Jesus won us new hearts (2 Corinthians 5:17), as well as the Holy Spirit, who lives in us (John 16:7). He also won us direct access to our Heavenly Father, so we can ask for the heart change we need—and he is joyful to give it (Matthew 7:7–8). All that to say: Whatever we give up is a pittance compared to the amazing riches and joy offered in God.

A gentle reminder: Unhealthy tech isn't just a gamer thing. Much of our tech is designed to usurp our attention and distract us into scrolling, viewing, buying, and playing more. In the world of Drool Tech (tech that has a mind of its own), searching for satisfaction is prevalent across all demographics and on all platforms. Some of us seek the feeling of importance and belonging. Others search for purpose. Still others look for a quick escape from reality.

Whatever your reason, my prayer is that this story will be one more little reminder to listen to that little voice in your heart, the one that tells you something needs to change. There is a God who loves you deeply, who has already done everything needed to give you a new heart, and who promises to never leave you and to use you to reach others who need hope for tomorrow.

In the coming chapters, we'll talk about the research regarding tech, how God has made our brains, and even get into the nitty-gritty of how to faithfully live out the gospel with tech, but for now let's keep the first things first:

Jesus himself went to the cross, not because he was worried God the Father might get mad at him, but he went "for the joy set before him" (Hebrews 12:2). He went in hope, knowing his death and resurrection would win victory for all who would believe in him. This is the mindset we take with our tech. We make our tech decisions in light of the joy set before us.

This is how we love God and use tech.

CHAPTER 2

What Do We Mean When We Say *Tech*?

Words matter. They shape thoughts, conversations, ideas, and ultimately, actions.

When we have a shared language, we are able to communicate effectively and easily. However, when we misuse and misunderstand words—even simple ones—things can go awry quickly.

My favorite example of this is from comedian Trevor Noah, a South African English speaker. In his comedy special *Son of Patricia,* he describes his first encounter with tacos: Upon eating them for the first time (a food not commonly found in South Africa, due to the lack of Mexican people), he was told to grab a few napkins because tacos are notoriously messy.

Trevor was horrified. "I need a napkin to eat this food?" he said. And while speakers of American English would say, "Of course! No big deal. Grab a whole stack of napkins." Trevor's repulsion at this concept came from different definitions of the same word. You see, in American English a napkin is a piece of cloth used to help clean small spills. In South African English, a napkin is a diaper used to catch a baby's waste.

You can see why tacos repulsed him.

When we talk about tech, I find it helpful to define the words we're using so we can have a shared understanding.

Two Types of Tech: Tool Tech and Drool Tech

This isn't a moral judgment of "good" or "bad" tech, but instead, it assesses the tech's design and intention. Tool Tech is designed to help you create. Drool Tech is designed to help you consume.

Let's start with Tool Tech. It's the digital equivalent of a shovel. You have a task, and Tool Tech extends your ability to accomplish that task more quickly and efficiently. Using a shovel is vastly superior to using your hands when you need to dig a large hole, and different shovels fit different needs. Have you ever tried shoveling snow with a spade or tilling your garden with a snow shovel? It does not go well.

Tool Tech operates at the pace of real life, extends your abilities, and requires an operator to function. If you stop working, so does the Tool Tech.

Drool Tech, on the other hand, may help you accomplish a goal like connecting with friends or enjoying your evening with a show or game, but it has goals of its own. Drool Tech is designed to take your time, focus, and money.

In this way, Drool Tech is the equivalent of a caffeinated drink or salted caramel. It provides you with something you want, but it comes with a hook that makes it more likely for you to want more. Drool Tech makes it more likely for you to stay longer than you intended, pay more than you planned, and come back more often than you wanted.

It also operates faster than real life; between the feeds,

goals, levels, and notifications, it is easy to get overstimulated. And if you stop working, Drool Tech keeps on going.

Now a little pop quiz. Which of these is Tool Tech and which is Drool Tech?

1. Microsoft Word
2. Netflix
3. Podcasts
4. Newsfeeds
5. Computers

We'll run through the answers one at a time—again, so we can be on the same page when we're discussing tech.

1. Microsoft Word (or Pages or Google Docs) = Tool Tech

Think about it: You've never been in bed, winding down from a long day, and received a notification from Microsoft Word saying, "Hey. Was thinking about you. I've noticed you haven't written anything for a while. I wanted to see if you might want to jot down a quick story—a haiku, maybe."

Word doesn't send you little notes saying, "Did you know Becky down the street has already started her book? Did you know it's better than yours? Here's a snippet to make sure you feel terrible about yours." This doesn't happen because Word is Tool Tech. It is a digital shovel created for writing. You can write, save, share, and edit with ease; when you stop, it stops.

2. Netflix = Drool Tech

Netflix is designed for attention retention. In 2017, the CEO of Netflix declared their main competition isn't other streaming services; it's sleep. They can only take so much time out of your work and family life until they need the only spare time you've got left.

And sure, you can watch documentaries on Netflix, and those can feel a lot like Tool Tech; except, few of us use the information we gain from documentaries to inform our actions or change our behaviors. It's what Neil Postman warned about in *Amusing Ourselves to Death*: the merging of reasoned information with entertainment would ruin discourse and free thinking in American society. He wasn't wrong.

Many of us come away from these films parroting instead of digesting and synthesizing information. After I watched *Supersize Me*, I decided I'd never eat at McDonald's again. Only Burger King for me! But it turned out, my newly inspired justice against fast food was more a fad revulsion against branding rather than any deep-rooted conviction regarding the quality of beef served. Can Netflix be used as Tool Tech? Sure. But its design and intention is to take your time, focus, and money.

3. Podcasts = Drool Tech (except for mine, of course)

Podcasts fall into a similar camp as the documentaries we're watching on Netflix. They can be Tool Tech if you're listening to multiple viewpoints on the same subject, and you're comparing, contrasting, and using your new findings to inform actions.

Chances are that you and I listen to what we like.

We choose podcasts that agree with our opinions, or ones that amuse us, and rarely stray into anything too challenging. Can it be done? Absolutely. I've learned and grown from the sermons of Tim Keller, *30 Minutes with the Perrys*, and *How I Built This with Guy Raz*.

The person creating the podcast is most definitely engaging Tool Tech. But the app delivery system mostly serves as Drool Tech.

4. Newsfeeds = Drool Tech

Your newsfeed is what you see when you open Facebook or X (RIP, Twitter). It might even be the Apple News page. In short, it's the flyby headlines or posts of the day, and they're designed to give you one of two things: The best news or the absolute worst.

To do this they need to know a little bit about you. Facebook, for example, looks at content posted by your FB friends, who and when the story was posted, whether you are likely to like, comment, or engage the post, and how likely others will be to engage that same information. Based on all that, they build you a custom feed of information that's tailored to your passions and interests.

The best news is the stuff you're most likely to retweet or share on Instagram or Facebook or Twitter, or the content you're highly likely to comment on when others post. The goal is user engagement. That's also true with the worst news. There's nothing like a good "Look at how *those people* are burning the world to the ground" headline to really fire up user engagement.

It's important to know many of the topics you'll see on your newsfeed are the same ones being covered by your favorite podcast. While there's a small chance that's the only news happening, it's more likely that this is the most salacious information available.

Emotions drive clicks and clicks drive revenue. Hence, your newsfeed is Drool Tech.

5. Computers = Tool and Drool Tech

Let me tell you a story to show you how a computer can be both Tool and Drool Tech. While teaching eighth grade Language Arts, I reserved every January to teach an especially mopey unit on tone and mood. There's nothing like arriving at school on a dark, rainy day to study how authors use words to bum you out.

We read *To Build a Fire* by Jack London, *The Little Match Girl* by Hans Christian Anderson, and my favorite, *The Tell Tale Heart* by Edgar Allen Poe, in which a guy hides the body of his roommate under the floorboards of their apartment. Now imagine you're one of these eighth graders, trying to type an essay on your school-issued laptop, all in your feelings, when your laptop sends you this notification:

"Did you know your laptop has Minecraft?" Simply click the banner and say goodbye to your feelings—and your assignment.

This is the trick with computers. And smartphones, for that matter. Sure, they can be a portal to learning, loaded with programs that certainly act as tools to encourage education, creativity, and work. And they are also platforms for enabling distraction.

There you go. Now we know what we're working with when we say *tech*. There are two kinds, and while each might have a screen or plug into a wall, they differ greatly in design. Tool Tech is designed to help you create. It supports your goals and extends your abilities to meet them. Drool Tech helps you consume. It distracts you from your goals, or extends them, asking you to give more of your time, focus, and money then you'd even intended.

If there are conflicts around tech in your home, there's a good chance they have to do with Drool Tech.

CHAPTER 3

Designed to Distract

How does Drool Tech work? How can a device, app, or streaming service be designed to take my time, focus, and money? Simple: Operant conditioning.

Operant conditioning is when a person or animal is trained in a new behavior through a process of rewards and punishment. It's the clinical version of the proverbial carrot and stick. Back in the 1940s, Dr. Donald Hebb coined the idea that neurons that fire together, wire together. This *Hebbian Theory of Learning* was a massive step in the study of human behavior because it opened the possibility that our choices might drive our change—practice might actually make permanent.

Today this idea is called *neural plasticity*. It's the idea that your brain adapts to your behavior, and that your mind can actually exert itself to influence the output of your brain. At this time in history the idea of people having a mind, yet alone a free will, was dismissed as "pre-scientific superstition" by some psychologists. One such psychologist was professor B. F. Skinner of Harvard University. While Dr. Hebb established the theories, Skinner established the practice of behavioral training.

Before we talk more about Professor Skinner, however, I

have one more character in the field of behavioral training to introduce: my grandma Audrey. Don't laugh. Grandma was no one to trifle with. She grew up in Kansas during the Dust Bowl and the Great Depression. That's the kind of childhood where you could freeze to death trying to use the outdoor bathroom in winter. On cold mornings she had to break the ice off her wash basin, in her bedroom. Grandma was tough as nails.

She did, however, have a soft spot for horses. Wild horses were caught and corralled on her family farm and she loved to recount how they convinced wild horses to take riders. The first way was her favorite and went something like this:

Each day, Grandma would walk past the horse's corral and set a slice of apple on the fence post. She wouldn't stay; she'd just set it there and walk off. After she left, the horse would warily nibble at the abandoned treat. It wasn't long before the horse connected the little lady walking past with the sugary morsels left on the fence post. Soon the horse was following her along the fence anytime she walked by.

As days progressed, Grandma would add steps to their routine: first it was treats from her hand; then she'd leave the treat and brush the horse while it ate. After several days the horse would take a blanket over its back for a few minutes at a time, then for the whole day.

As time went on the horse would take not only a blanket but a saddle, cinched down, with a rider. By the time Grandma was on the horse's back she didn't need a bit or bridle because the horse went willingly wherever she wanted it to go. It's easy to see why she had such fond memories of this apple/carrot version of operant conditioning. There was, however, one major drawback: time. This was a fine way to train a pet, but a rough way to make a profit.

That's where the second method comes in. This second

method was all stick, no carrot. It involved a young person, almost invariably a man, who would throw himself on the back of an angry young stallion and try to hold on for as long as possible. People would place bets on how long he'd last, or how long until he died in the process, and riders would continue to jump on the horse until the horse was so tired it simply couldn't buck any more. Thus broken, they would toss a bit and bridle on the horse and it would take any number of riders without protest. This stick method was crueler, sure; but it was far more profitable.

This brings us back to Professor Skinner. Around the time my grandma was raising her horses in Kansas, Skinner was changing the world of behavioral psychology. Professor Skinner used this same carrot and stick method to train rats to press a lever on command. Initially he did this training with rewards only. A light would go on, a sound would activate, and if the rat pressed the lever it would get a treat. This worked until the rat didn't want the reward. The quickest solution was to electrify the floor. The rat flailed around until it hit the lever, which turned off the electricity. The rat learned quickly: Light goes on, you push the lever whether you want a treat or not.

In Skinner's thinking this verified his premise: It is the environment, not the mind, that determines behavior. His predecessor John B. Watson argued there is "no dividing line between man and brute," and Skinner believed he'd proved it.

Instead, we now know his work to have unwittingly progressed the very scientific field he held in such contempt: the Cognitive sciences. Using his methods we have learned that brains are malleable, they change with our choices, and current choices influence future behavior. Jeffrey Schwartz, a research psychiatrist at the School of Medicine at the University of California at Los Angeles and one of the world's leading experts in the field of neuroplasticity says, "*Neuroplasticity* refers to the ability of neurons to forge new connections, to blaze new paths through the cortex, even to assume new roles. In shorthand, neuroplasticity means rewiring of the brain." Simply put: the mind matters.

To understand why neuroplasticity is key to how Drool Tech works, picture your brain as a field. The more you walk a path, the more pronounced and permanent the path becomes. This is the case in a study of London cab drivers that found the rigorous cab driver test, known as *The Knowledge*, required such demanding spatial reasoning that veteran cab drivers developed a much larger hippocampus than their general population counterparts. In fact, the size of the hippocampus matched the length of the driving career. Given the city of London encompasses more than six hundred square miles, memorizing such massive spatial details gives us insight into how the human brain adapts to incredible demands.

It's not just cabbies whose intentional choices create permanent brain change. Accounts of this power are replete in the medical field as well. Dr. Norman Doidge's *The Brain that Changes Itself* highlights the magnitude of what neuroplasticity can mean for medicine.

One such account is of Cheryl Schiltz, a woman suffering from debilitating vertigo. She lost her job, her independence, and couldn't even lie down without feeling as though she were falling through the floor. Her condition was believed to be untreatable because her vestibular system, the part of the body responsible for balance, was destroyed as a side effect of surgery medication. Her doctors told her that even if they had a surgery that could help, there wasn't anything left to fix.

Then Cheryl met Paul Bach-y-Rita, a doctor who suggested an experimental therapy: electrodes on her tongue. Dr. Bach-y-Rita used a strip of plastic equipped with 144 electrodes that would stipple her tongue as she moved her head. Her first session gained her minutes of independent standing after five years of crippling dizziness. Not only could she stand while using the device, but a residual effect occurred: After twenty minutes with the device, Cheryl could stand for another seven minutes. Over the course of the next year she took the device home and practiced. She trained until she no longer needed the device, and she no longer suffers from vertigo.

That, however, is just the power of repeated choices. It's also why behaviors that involve reward hormones or neurotransmitters, like dopamine, can be so habit forming. These are like walking through a field with a flamethrower. The hormones help scorch a more permanent path into our brains, making these choices a quicker default in the future. This is why Drool Tech can be so unhealthy, especially for young developing brains.

How does drool tech train us?

In his book *Hooked*, Nir Eyal lays out the basics of behavioral training. There are four basic components:

1. Trigger: It can be as simple as boredom or as intentional as a notification.
2. Action: This means practice, and practice creates neural pathways.
3. Reward: You'll get feedback in some way. A like, comment, level, medal, loot, or burst of updates.
4. Investment: You've got skin in the game. You create content and therefore, *you* are the content. People don't just like your post, they like you.

Drool Tech's job, then, isn't to create something you need

or even something you want; its job is to keep you coming back for more. When the choice becomes a habit, you become a core user of that product, and you'll eventually be convinced you need something you didn't even want—and perhaps lose the ability to choose it at all.

Choosing Wants Over Needs

Consider Snapchat for a minute. Ask any teen who uses it, "Have you ever kept a Streak alive when you should have been eating, sleeping, hanging out with friends, or doing school work?" The answer is always "yes."

Carefully review that list again:

- Relationships
- Food
- Study
- Sleep

That list, however, is Maslow's hierarchy of needs. You need food and sleep for survival; you need relationships for purpose and hope. These are basic human requirements, and we can be trained to give them up for something we never even wanted. My friend Cameron runs an addiction recovery ministry, and one day, I asked him how he recognizes addiction. His answer surprised me with its simplicity: Addiction is any time you give up what you need to get what you want. In that simple answer we have a way to see whether our behavior is healthy, compulsive, or tipping into a lack of control that requires outside care and support.

But why does behavioral design matter?

It's one thing to understand this happens, but why should we care? If I enjoy the product, if I like the game, social media platform, streaming service, or news outlet, why does it matter that I'm being trained? Behavioral design matters because we become convinced we need things we don't want.

Again, let's examine Snapchat. They designed a messaging app that tracks your consecutive streaks with friends. You send a message, or a "snap," to a friend. They snap you back within twenty-four hours. You now have a streak of one. Keep snapping daily and your streak grows. Miss a day and your streak dies, potentially taking your friendship with it.

No one asked for this feature. There was no flood of Instagram reviews saying, "We love Direct Message, but what we really want is an app that keeps score." Yet Snap Streaks drove user engagement, and that drove revenue.

What about YouTube? It has more than 2.5 billion monthly users, and is a master of eeking out an extra minute. If YouTube can get just 10% of users to stay just one extra minute, it gains more than 170,000 days worth of additional viewing —that's equivalent to 465 years, or watching all ten seasons of Friends 47,000 times. Keep in mind, seventy percent of YouTube's viewed content comes from their suggestion bar. Apparently I'm not the only one going to YouTube for DIY help and staying for those trick shots (thanks for nothing, Dude Perfect). The longer I stay, the more money they make.

So Snapchat drives engagement with emotional manipulation, YouTube leverages curiosity and an elegant algorithm feed, and then there's TikTok. TikTok goes all in on quantity over quality. You don't have to pick a topic, channel, or even a specific video. Just scroll your feed and new videos populate. The more you pause, or scroll back to previous videos, the more of that content shows up. While the videos only last about thirty seconds, the average user watches roughly ninety minutes of TikTok content each day. That repeated training of users pays off for the time/focus/money goal of social media— fifty percent of TikTokers say videos longer than a minute make them feel stressed, which means about one third watch videos on double speed and they return to the platform because longer-form mediums are anxiety inducing.

And in all this, we're not even diving into the way suggested content might influence our thoughts, decisions, buying habits, or votes. The content suggested on YouTube for me to watch and remember is specifically cultivated to get me to keep viewing.

The Greatest Concern

The greatest concern is these behavioral designs have no moderator. Ask a design team, "How much is "enough" engagement?" The answer, inevitably, is "more."

The engagement process is less like adding sugar, or even caffeine, to food and beverage products and more like Coca-Cola circa 1900. Early on, Coca-Cola was a delightful new soft drink which, amongst other ingredients, included extracts of cane, Kola nut, and Coca leaf, or as we know them: sugar, caffeine, and cocaine. Coca-Cola's engagement strategy was "more." Eventually the U.S. government stepped in and said cocaine wasn't allowed anymore. The product was less engaging without cocaine, but it was also less destructive.

Drool Tech is designed differently from other tech— it wants to take your time, focus, and money. The design is based on the idea that we form habits around thousands of little choices so, if a game, show, or device can trigger little actions, they'll build little investments that can make a person feel like they need something they never even wanted.

It's often not enough to just tell our kids "Not now" for a smartphone or a particular app, but when we can better know and then explain the heartbeat behind our parenting decision, we are able to establish both rationale and intent. Children are not fools; coming to them in good faith, equipped to discuss how the science of technology design supports or hinders our Christian walk allows us a humble entrance to the conversation. When it comes to taking action, there is much we can do to mitigate Drool Tech's negative impacts.

Part 2: Dealing with Digital Design

CHAPTER 4

Your Kid Doesn't Need a Smartphone

I thought about making this the world's shortest chapter: "Does my child need a smartphone?"

"No."

I tried that once in a presentation. It didn't end well. I thought it'd be a funny way to introduce a topic that can feel loaded with all kinds of pressure, expectations, opinions, and teenage drama. It turned out to be one of those times an idea is funnier in my head. My hope for begrudging laughter was shattered by a silence so complete it could only be described as obstinate. I should've seen it coming.

As parents we don't just want the right answers. In fact, I'm continually impressed with how many parents already know the right answers. What parents want is help putting the right answers into practice. It's one thing to know a smartphone is a bad fit for your child and quite another to face that child and explain your logic without sounding like an old man griping about "kids these days."

So, with that in mind: No, your child doesn't need a smartphone.

In fact, they don't need any of the personal devices you might have lying around your home including laptops, tablets (Kindle Fires, iPads), etc. This group should by and large be operating on shared family devices located in public spaces. Here's why:

Our goal as parents is to raise healthy kids who love God and use tech. Too often, tech is the default setting. It's just so easy to hand over a screen. But if we want our children to reach their full potential, to live out the calling God has on their lives, and to launch successfully into the world as young adults, well, we must parent on purpose.

Our personal family goals revolve around knowing and loving God, serving others, and being kind and curious students (and grownups). As parents, we are working to discern the gifts each of our children has been given and to cultivate them. For example, our oldest son is methodical, organized, and athletic (and ten). We watched the local high school basketball team play this winter, and Owen asked for a stat sheet so he could tally shots, rebounds, and assists. I resisted telling him that this might be perceived as a tad bit nerdy and instead, readied him with a clipboard and pen.

It's these real-life experiences we want for our kids—chances for them to grow into who God has called them to be. And often, time online flat out impedes that growth.

With that in mind, here are my top five reasons to delay giving your zero- to sixteen-year-old a personal device:

1. Phones are really just app delivery systems.

Let's be real. When your child asks for a phone, he or she is asking for certain apps.

Most aren't dying to be on Google maps or check the family calendar. Most apps are Drool Tech designed for consumption. They are crafted to take your time, focus, and money, and are dedicated to cultivating habits beneficial for their revenue. Because children don't have fully developed frontal cortices, they are not able to inhibit behavior well (and this doesn't fully kick in until age twenty-five!) When an app makes a suggestion, there's a good chance your child will follow down the rabbit hole. They can't see any other choice.

2. The pitfalls of a smartphone do not outweigh the positives.

Smartphones come with a host of purported positives: connection with friends, apps that help you learn, the ability to set an alarm on this fancy $1,000 clock. While your son or daughter is asking for such resources it's important to recognize that the downsides of smartphones, specifically, outweigh the benefit for young, developing brains. Just look at focus, for example. In his recent article titled "Get Phones Out of Schools Now", NYU professor Jonathan Haidt notes that "heavy phone or social-media use may also have a cumulative, enduring, and deleterious effect on adolescents' abilities to focus and apply themselves."

It's not just a theoretical risk of lost opportunity or 'kids these days' and fear of what we as adults may not understand. The harm posed by constant access to the internet is measurable and real. A 2019 study showed that social media was between two to six times more likely to cause harm than watching TV or playing video games. Researchers found that youth who spend "more than two hours on the internet are forty-eight percent more likely to self- harm than those spending less than two hours on the internet." It actually gets worse. "Those spending more than seven hours on the internet are more than twice as likely to self-harm than those spending between thirty minutes and one hour on the internet" each day.

There is plenty of conjecture as to why this is the case, but at least some of it is due to the sheer barrage of content. The short, frenetic snippets of social media lead to a frenzied, often overwhelmed state of mind where focus becomes impossible. "Directed attention is the ability to inhibit distractions and sustain attention and to shift attention appropriately," says Dr. Michael Manos of Cleveland Clinic Children's. "If kids' brains become accustomed to constant changes, the brain finds it difficult to adapt to a non-digital activity where things don't move quite as fast."

At the time of writing, TikTok is one of the top apps among teens, with fifty-eight percent of teens visiting daily. Videos on the platform average between thirty-two to forty-three seconds, with TikTok users averaging ninety-five minutes of viewing each day. That's between one hundred twenty and one hundred eighty videos, ranging from sports, kittens, and friends to war, politics, self-harm, pornography, disordered eating, and advice on mental health that may or may not be vetted by a professional.

Smartphones have some benefits, but their downsides simply outweigh the potential perks of giving them to children and teenagers who are still learning how to learn, focus, and be healthy.

3. Bullies and strangers loom large in online spaces.

I want to strike the right balance here. The goal isn't to lock your child in your house, keep him or her by your side, or always be looking over your shoulder for a kidnapper (like many of us raised in the '90s were doing). The goal is to foster opportunities for a "real world" childhood to occur. But the reality is this: When we send our children into places like Roblox unattended, we are setting them up to encounter danger.

If we are concerned about the mental, emotional, and spiritual development of our children, then we need to honestly ask ourselves how giving them a smartphone aids them. The research is clear that the more time young children spend online, the more likely they are to encounter bullies and ill-intentioned strangers. And yes, we need to coach our children how to interact with others when they're mean and hurtful; but we can also set them up for success. Less time on social media and the internet in general is a great starting place.

4. Guarding hearts and eyes begins with us.

The average child sees his or her first pornographic image at age nine. It's not really a matter of if our kids will encounter porn, but when, and we want to be ready with more than good rules and thorough punishments. As writer and mother Greta Eskridge says, making them porn resilient. By giving our children distance from the constant call of the internet we give them freedom from that siren's call to distraction and allow them to enjoy their childhood while minimizing their risk of unhealthy exposure to unsafe content.

When we hand our children a smartphone, we need to be sure we're having conversations around this topic. If your child isn't ready to hear the words pornography and predator come out of your mouth, then your child isn't ready for a smartphone.

My favorite resource here is Good Picture, Bad Picture, as it provides an easy way to enter the conversation with your child and serves as a script for navigating this sensitive area. (See the Appendix for additional resources).

5. Our minds aren't designed for constant entertainment.

Finally, the last reason to delay handing your child a personal phone is the entertainment factor. The teacher in me can't help but notice the rather startling shift in how we have to educate our kids simply because they've been trained from a very young age to expect entertainment.

Rather than building self-control, patience, or kindness while waiting at dinner or the grocery store, children are often handed a game or show. While it might buy a few minutes of peace, we must go back to the beginning: Every decision we make as parents needs to reflect our end-game goals.

Phones are full-blown gaming platforms. When students have video games in their pockets all day, it's (justifiably) hard to focus on school, sports, and even other people. Why would I listen to a lecture on Odysseus when I could be playing Fortnite?

Now, at some point, you and your child might reach a spot where he or she is ready for independence in this area. Here's a quick litmus test to assess readiness:

Is your child over the age of fifteen?

Remember, our goal as parents is to raise healthy kids who love God and use tech. That starts with us giving our kids a chance for a childhood, and a chance to learn, grow, and develop in a way that encourages and challenges them. Discipleship, I've found, rarely takes place when we're consuming tech. Waiting until sixteen to hand over a smartphone allows your child a chance to develop more fully into the person he or she has been designed to be.

Is your child trustworthy?

Assess the little things. How does he or she manage school, friends, family, sleep, and money? Luke 16:10 tells us that we need to be faithful with the little to be trusted with the big. In the same way we would expect our child to complete Drivers Education classes before handing over the car, consider the training you might require before handing over a phone: Practice with shared family devices or perhaps a dumb phone before handing over the information highway that is an iPhone. (See the Appendix for suggestions.)

Is your child asking for a smartphone?

Generally, kids aren't going on giant tech benders if they're not given phones before the age of eighteen. What is happening instead is that young people are opting out of smartphones all together. If your child is of an appropriate age and has proven trustworthy, and there is a legitimate need (beyond "everyone else has one!"), then begin having the conversation around wise smartphone uses.

If I could leave you with one takeaway here it's this: Smartphones pose more harm than good. When we parent with the end in mind—and have clear goals for how we want to steward the gifts of our children—we can more clearly assess the role that technology plays on the minds and hearts of our children. So no, your child doesn't need a smartphone.

CHAPTER 5

Social Media is a Net Loss for Youth

I didn't want to write this chapter. In fact, over the past four years of giving live workshops I've never directly addressed social media. My reasoning is simple: Social media sucks the air out of the room. In a given workshop there may be a dozen topics to address but a single question about social media can take all the time and focus we have while leaving very few usable solutions. It's a hydra with a thousand heads, and answering one question leaves three more in its place.

Here's why a simple social media conversation can get convoluted quickly:

Every social media experience is different. Even popular apps like Snapchat and YouTube vary widely. This means one question about app A needs a different, and specific, answer for apps B–Z.

User choice often matters more than the app itself. A wise person on TikTok is better off than a fool on Facebook. Parents come with understandably specific questions on how to fix a problem they see with their child. Yet many times what they want is app-based control to ensure their child makes better choices, while what they need is a relational process to talk tech and assess health, along with practical tools to the unhealthy tech.

Tech changes rapidly. To make matters worse, tech tips are often not interchangeable or transferable. All of your LinkedIn knowledge won't help much in navigating Instagram or Discord. And by the time this goes to press, "X" will be back to Twitter (Lord willing) and teens will have moved on to some new form of communication and distraction.

So while I'd rather not address it directly the fact remains that social media persists, parents have questions, and there is too much evidence declaring the dangers of social media to stay silent. This section will be broken into four parts:

- We will begin by looking at social media as the new digital cul-de-sac.
- We will consider what makes social media the ultimate Drool Tech.
- This will lead us to the negative impacts of social media on the mental health of youth.
- And finally, we will survey practical steps we can take to navigate this conversation.

Let's begin.

1. Social media is the new cul-de-sac.

My wife grew up in a 1990s suburban cul-de-sac, full of neighborhood kids, block parties, and late night games of flashlight tag. She spent large parts of her childhood in and out of neighbors' homes and backyards, riding bikes up to the perimeter of the larger neighborhood, and learning to negotiate peace treaties with other kids.

Contrast this nostalgic image with our current neighborhood: Also suburban, also a cul-de-sac, also (theoretically) full of children—there is an entire elementary school dedicated to our subdivision. I know there are kids hiding in these homes. Yet our street is quiet and the streets around us are quiet. We occasionally see a few boys shooting hoops or a mom dragging toddlers to the playground, but overall, the streets are not filled with the sounds of children playing (unless you count the sounds of our own children screaming wildly as they wrestle on the trampoline).

We've had neighbors stop us, concerned, when we let our children walk alone to the mailbox (which I can see from my front door); others asked if our children could attach six-foot flags to the backs of their scooters so cars could see them more easily. Did I mention we live in a double cul-de-sac and the only people who drive in live here? It is hard to imagine a safer physical place for children to play outside, and yet—they're all inside.

It's actually not surprising. Parents have become more fearful of the world and therefore more likely to send their children into "safer" digital spaces. Psychologist Jonathan Haidt writes, "parents (on average) say children should be at least 10 years old to play unsupervised in their own front yard. They say that kids should be at least 12 years old before being allowed to stay alone in their own home unsupervised for one hour. They say kids should be 14 before being allowed to go, unsupervised, to a public park." These are the same parents who were allowed to do these activities by the age of eight.

This transition to digital spaces made the internet the new cul-de-sac: it's where kids go to hangout, make friends, and have fun. Teenagers spend eight and a half hours a day on average consuming entertainment tech, and one of the major forms they consume is social media. While it's important to have friends, belong to a close knit community, and stay in touch, let's be clear about where this digital cul-de-sac is: It's by the freeway, and a federal penitentiary, in the red light district, at midnight. When we keep our children indoors and online in the name of "safety," we must recognize the irony of the claim we are making.

We want to nurture childhood. We want the resiliency and fortitude that comes from skinned knees, ghost-runner baseball games, and snowball fights. The longer we delay their entrance to social media, the larger their chance of forming healthy habits that will serve them well when they finally do have to make social decisions in a digital playground.

2. Social Media is the ultimate Drool Tech.

Drool Tech is built for attention retention. The behavioral design of social media is intended to extract as much time as possible from the ever-shrinking margins of your attention in the digital world.

These reduced margins are what caused the U.S. Surgeon General to warn Americans that social media uses a host of tricks to make it difficult for users to walk away. The advisory cautions that "social media platforms are often designed to maximize user engagement, which has the potential to encourage excessive use and behavioral dysregulation" and points to engagement design including "push notifications, autoplay, infinite scroll, quantifying and displaying popularity (i.e., 'likes'), and algorithms," which it says could (and does) create compulsive behaviors.

This isn't a surprise to Tristan Harris, co-founder of the Centre for Humane Technology (CHT) and former Google Ethicist. Such design mechanics are part of what he calls the digital "race to the bottom of the Brainstem." Harris worries that modern society is conditioning youth to have tech co-dependence where adversity is treated with a "digital pacifier" leaving swarms of young people unable to cope with important, and difficult, trials of real world adulthood.

This is no small issue. Nearly ninety-five of U.S. teens use social media on a regular basis. As the body of research in the field of social media and youth continues to grow, so does concern. Dr. Mark Griffiths of Nottingham Trent University found social media companies use a host of tricks including unpredictable rewards, social affirmation, haptic feedback (your phone buzzing), social connection, the fear of missing out, and competition to encourage users to return. It seems they work. Nearly one third of social media use is due to people staying on social media longer than they intended, while one third of girls ages eleven to fifteen say they feel "addicted" to a social media platform; this number tracks with a survey of eighth through tenth graders who average three and a half hours a day on social media.

With a third of teens saying they use social media "almost constantly," there is little doubt that social media is doing a great job at engagement. This is effective for business, but this investment is not without cost. Our attention is our most valuable resource, and it's limited. As Christians, we are called to live wisely, stewarding our lives unto the glory of God (Ephesians 4:22–24, 5:15–17; 2 Corinthians 5:17; Mark 12:30). Social media operates largely as Drool Tech, is designed to keep users engaged beyond their intention, and ultimately, competes for user attention in a way teens especially cannot combat on their own.

3. Social media negatively impacts mental health in young adults.

If points one and two weren't enough for you to press pause on social media, let me join the chorus of tech ethicists, teachers, doctors, pastors, and kids themselves who are shouting to anyone who will listen: America's youth is suffering a mental health crisis.

In 2021 nearly three out of five teenage girls in the United States felt persistently sad or hopeless. While the stats on boys aren't quite as severe, it's a stretch to say they're better: one third of teen boys reported feeling the same persistent sadness or hopelessness. Those are numbers we've never seen before in youth mental health; but if you parented teens through the pandemic, you've likely had a front row seat to the mental health crisis occurring. The research is not opaque here: We need look no further than social media for our culprit.

What sets social media apart from the straw men of previous generations (Think: rock and roll, the war on drugs, MTV, etc.) is the overwhelming body of research and evidence-based action plans that the social media spiral has on its side. While parents might have had moral qualms with the trends of previous eras, for iGen, as dubbed by Dr. Jean Twenge, the growing body of research is clear and concise: the negative impacts of social media on youth is a clarion call to stop prioritizing profits over the health and wellbeing of children.

Leading the charge is Jonathan Haidt, professor at New York University (NYU) and outspoken advocate for immediate and intentional intervention and legislation on behalf of youth. His article Social Media is a Major Cause of the Mental Illness Epidemic in Teen Girls minces no words: "There is now a great deal of evidence that social media is a substantial cause, not just a tiny correlate, of depression and anxiety, and therefore of behaviors related to depression and anxiety, including self-harm and suicide." He is taking that same passion, by the way, to extend best practices shown in research and get smartphones removed from schools—but that's a conversation for another day.

America's surgeon general, Vice Admiral Vikek Murthy, compiled 104 of the most prevalent studies, models, and meta-analyses available when it comes to the impact of social media on the mental health of youth. The end result of his findings? "Personally, based on the data I've seen, I believe thirteen is too early…the skewed and often distorted environment of social media often does a disservice to many children," he declared.

This is a formative time of development for young brains, which is why social media impacts children more greatly than adults who may engage it for the same amount of time. One reason for this is that part of brain development involves taking risks. When taking risks outdoors we often get obvious scars (those tooth marks through your bottom lip, that gash on your elbow), but the damage incurred from internet exploration isn't so obvious. And because this is all happening at the developmental stage of life when depression and other mental health challenges are prone to arrive anyway, the introduction of social media to youth makes for a devastating combination.

Additionally, an overuse of social media is linked to a decrease in user happiness. In 2022 Braghieri et al tried a unique approach to assess the toll social media has on mental health and user happiness: They analyzed the mental health data of university students during Facebook's staggered rollout across the United States. As Facebook was introduced to universities researchers found "an increase in depression (9% over baseline) and anxiety (12% over baseline) among college-aged youth." The introduction of even rudimentary social media from the early 2000s caused a marked decline in happiness and mental wellbeing in young adults.

Keep in mind, this was just the introduction of Facebook. Researchers like Jonathan Haidt and Jean Twenge note that with the introduction of Instagram these problems became much, much worse, especially for girls. They don't mince words, "There is one giant, obvious, international, and gendered cause (for this mental health decline): Social media… By 2015, it was becoming normal for twelve-year-old girls to spend hours each day taking selfies, editing selfies, and posting them for friends, enemies, and strangers to comment on, while also spending hours each day scrolling through photos of other girls and fabulously wealthy female celebrities with (seemingly) vastly superior bodies and lives. The hours girls spent each day on Instagram were taken from sleep, exercise, and time with friends and family. What did we think would happen to them?"

Not long after its inception social media pivoted from connecting people to leveraging the human desire to be loved and known, and the fear of missing out, to build an empire around making users feel the need return in order to be a part of something, what we now call FOMO, or the fear of missing out. Researchers have found that leveraging FOMO for engagement is "associated with depression, anxiety, and neuroticism." It's not surprising then that social media is such a force for mental degradation, especially in teen girls.

I could go on, but the point stands: much of social media is designed to capitalize on developing minds in a way that is inherently unhealthy and unsafe. When I queried Google Scholar for ["social media" depression], it yielded 1.3 million results. Try for yourself. For brevity's sake, I'll include in the appendix books that have informed my conclusions as well as shaped the way we talk about tech use in our family.

The point is this: Social media, perhaps most out of any technology, is not a fair fight, especially for our children. Some of the most poignant data points exposing social media for what it is are from within the walls of the high towers itself. In 2021 a report by the Wall Street Journal exposed internal research and training documents at Instagram showing they knew their product was harming their main clientele. One slide read, "Thirty-two percent of teen girls said that when they felt bad about their bodies, Instagram made them feel worse." This was at a time when more than forty percent of Instagram's user base was aged twenty-two or younger.

Finally, let's circle back to that usage stat for eighth to tenth graders: twenty-five percent are spending more than five hours a day on social media. But a 2017 study found that a mere two hours a day increases the likelihood of depression and suicidal ideation by twenty percent, increasing to sixty-six percent increase when kids are using it for more than five hours a day. This means for at least a quarter of our teens, social media isn't just messing with their minds; it's threatening their lives.

I know this is heavy and heartbreaking. If you're parenting teens, you've likely seen the effect social media has on them and their peers, changing the children you knew into teenagers who seem unrecognizable. As a former teacher and Young Life leader (and uncle to seven teens), I have watched social media take hold of sweet kids, leaving them shells of their former selves.

Before we get to the tangible steps we can take to make social media an even fight, let me remind you of this:

All hope is not lost. Parents (or loving grownups!), you

are still the number one influence in your child's life. Your prayers, your influence, and your intervention are still relevant. Remember, our stories are long and often windy, and our walks with Jesus aren't always—in fact, hardly ever—a straight walk up and to the right. They, instead, like Christian in Pilgrim's Progress, are filled with obstacles, temptations, setbacks, all revealing the greater hope we have in heaven than here on earth. Your job, then, is to remind your child of to whom they belong and the hope they have in Christ.

4. Practical steps we can take to navigate social media

Let's talk now about the tangible steps we can take to make this a more even fight. Despite the overwhelming research, we are not helpless here. We can take simple steps to protect our children and to support those who are fighting for our families.

First, begin the conversation. The goal of this book is to equip you to have the tricky conversations surrounding tech. I've found that children as young as four can begin to understand the differences between Tool and Drool Tech. This doesn't have to be a one and done talk, but can be something you come back to time after time. (My elementary age sons can spot Drool Tech a mile away).

As we'll discuss more in chapter 8, we can reflect on whether our tech is healthy by whether it improves or impedes our relationships, emotions, sleep, enjoyment, and time. Your child is going to follow the example you model for them, and in this case, "raising them up in the way they should go" means humbly receiving feedback, modeling reflection on your own tech, and then making changes when needed. Finally, set up safeguards around your devices, network, and family. For additional support, check out the Family Tech Framework in the Appendix.

Second, use only the social media that is beneficial to your analog life. Here's what I mean: If it helps you achieve your goals or it builds on your real-world relationships, great. Keep using it. If not, cut it. You'll know it's in its proper place when it waits for you. Try checking it at a specific time, on a specific day, and then walk away. That could be every day at a specific time.

If you have to use social media for more than thirty minutes a day, you're doing it wrong. Connecting with people, checking details, and sending personal messages can be done in surprisingly little time if we put our minds to it. Maybe you're like John Mark Comer; you've written the book on getting rid of hurry and you only check it two days a week. Maybe you're author and professor Cal Newport and you don't have a social media presence at all.

And finally, delay, delay, delay. Perhaps the most simple step to take here is to delay when we allow our children to enter the world of social media. While there might be a national push to delay smartphones until eighth grade, remember the words of the Surgeon General, who declared that thirteen is too early for children to be on social media. The benefits of access simply do not outweigh the risks.

Your child does not need social media to be well, and you will not hurt your child by waiting to give them personal access to the digital cul-de-sac until they prove they are safe, healthy, and trustworthy. The goal here is not avoiding technology; it's emphasizing real-world friendships, real-world fun, and learning to be present in an already challenging life unfettered by digital distractions, devices, and apps.

CHAPTER 6

What Do We Do About Video Games?

As I shared in my tech testimony, when I was eight my dad brought home a Nintendo Entertainment System and my life was changed forever.

Now, from what I understand, other families (ahem, my wife's family) also brought these gaming systems home, and their children played Mario Brothers and Duck Hunt happily and sporadically, and like LEGO and Barbie, grew out of them. The consoles were sold or given away and these other children of the '90s look back on the games with fondness, but not obsession, like me.

These games did something different in my brain: they opened a portal to an achievable adventure where from Friday night to Saturday night (my brief window I was even allowed to play), I was the hero.

My point is this: Some of you did grow out of games and it's hard for you to understand the obsessive way others of us think and talk about gaming. In the early '90s, everyone assumed kids *would* grow out of games. But here we are in 2024, and the average gamer is thirty-two years old. In fact, in the United States, there are more gamers over the age of forty-five than there are under the age of eighteen. That means all of us don't just "grow out of it."

We are hooked.

Here's what I want you to know:

Video games are awesome. There are trillion dollar companies throwing billions of dollars at some of the most talented artists, storytellers, game designers, and hardware developers on the planet. The games they make are incredible–but that doesn't necessarily mean we should jump in without looking first.

We don't grow out of it. Just look at the history of new gaming consoles. You take a kid who has never played video games and put her on the original Nintendo, it'll blow her mind. We didn't make the Super Nintendo because the original wasn't cool anymore–we made it because we were bored. After the Super Nintendo we made the N64, then the Playstation, the Xbox, and onward through the Xbox One X. We now have a device you strap to your head to project the game directly into your skull. Basically if your head doesn't smoke while playing it, it's not intense enough. Your average adult gamer has to have something that's intense enough to cut through thirty-plus years of entertainment scar tissue. This also helps explain why our children struggle to sleep after playing modern games. What took my generation decades to build up to has become the baseline for younger gamers.

If gaming habits are unhealthy, change will only happen on purpose.

So how do we decide if our children should play games at all? And if they play, which games should they play? For a deep dive on picking games well, see the Appendix, where I've included my resource on video game assessment—this will help you if you're wondering about a specific game.

When we're talking about if and when our family plays video games, there are six key questions to consider. Again, these questions are to frame a conversation (especially with older children, ages six to sixteen) that centers on living in intentional pursuit of who God has called us and our children to be. That's our focus—raising young people who love God and use tech.

1. Does the protagonist pass the dinner test?

Ask yourself a simple question: Would you have this character over for dinner? Your child is spending a lot of time with this character, and if his or her behavior, language, or values wouldn't make for a good friend, they probably don't make a good digital or online companion. While I might be interested to hear how Mario and Luigi started out as plumbers and became the saviors of a magical mushroom kingdom, I am less interested in having the cast of Grand Theft Auto in my son's bedroom for any amount of time.

As my literature professor Suzanne Wolfe put it, just because it isn't real doesn't mean it isn't true. If the main character of this game doesn't pass the dinner test, then we must reevaluate why we're playing the game.

2. Is violence a part of the game or the point of it?

When Mario jumps on one of those little brown goombas it goes flat. Maybe it dies? It's safe to assume it had a family, but that's not the point of the game. You move on with your life, jump on the flag pole, and save the princess from the turtle/dragon/dinosaur. But that violence is quite different from the spine-ripping sort made famous by games like Mortal Kombat. If you're not familiar with it, and want to lose sleep tonight, just Google "Mortal Kombat Fatalities" and watch whatever video comes up first.

As a family, set a clear standard for the level of violence deemed acceptable. Culturally, our threshold for gore is quite high—we need to consider how what we view in games and online impacts the condition of our hearts, as instructed in Proverbs 4:23.

For some families, a game like Fortnite is a murder simulator akin to *The Hunger Games*; for others, it's a harmless game of laser tag played in a digital world. We must ask the question and establish a baseline. If you're not sure what the violence is like in a particular game, check out the online resources from the Entertainment Software Review Board (ESRB.org). This is the group who puts the letter grade on the outside of the box (E for Everyone, T for Teen) and their site does a great job giving quick game overviews that are clear, concise, and unbiased. So far, ESRB has remained an observer of the gaming world, not a marketing branch of big gaming. If you still want to know more, jump on YouTube: search the game name followed by "game play." A few minutes of this kind of game review will give you an excellent feel for whether violence is the point of the game, or just part of the greater experience.

3. Is the language of the game helpful or hurtful?

If a game is played live (Fortnite, Overwatch, or League of Legends) there is a high likelihood that the language your child is hearing by other players is unhelpful–or potentially harmful or hurtful.

Again, I recommend ESRB.org for an assessment on language. Words matter—our own and the ones we hear. Ephesians 4:29 instructs us to "Let no corrupting talk come out of our mouths, but only such as is good for building up, as fits the occasion, that it may give grace to those who hear." Video games are not exempt.

4. Can the game be beaten?

Humans have played games as long as there have been societies. Archeologists have found board games dating back more than five millennia. Whether ancient or modern, all games have three things in common:

- There's a challenge: You know what you're supposed to do. In the ancient game of Mancala, capture all the rocks.
- Clear rules are established: Players agree on the method of competition. One rock per hole. One person goes at a time.
- Finally, there's a victory condition: You know when you've won (or lost). When it comes to Mancala, typically it's the player with the most rocks who wins.

Remove any one of these elements and we no longer have a game. Maybe a fun pastime or an odd challenge is left, but the game is gone. This is where many video games run into trouble. They've moved from a competitive pastime to engagement traps.

World of Warcraft (WoW) is one of the best examples of this entrapment. In 2022, WoW had about 1.175 million players daily, playing an average of 3.7 hours each day. Since it came out in 2004, players have invested more than 3,258,720 *years* of playtime into this game. Once after a talk a man came up to me with two incredible facts about his experience with this game specifically: He had lowered his play time to just thirty hours a week, and that the game didn't start until you got your character to level 59/60. Up until that point, he said, the game was just a tutorial.

That's the definition of "hard to put down." Contrast this to Mario Brothers, which has all three game elements yet takes an entirety of two hours to beat if you play it straight through. When we live with Christ at the center of our lives, that includes our time. If we decide to play video games, we must take into account the amount of time we are investing in them.

5. Do the themes of the game align with biblical truths?

In the same way that books and movies portray powerful themes, video games are designed to promote agendas. Many of the most popular games for young and old alike include messages about religion, purpose, survival, gender, sexuality, politics, and mental health.

Before playing a game, assess the key themes presented: Do they align with the Bible? Philippians 4:8 instructs us to think on "whatever is true, whatever is honorable, whatever is just, whatever is pure, whatever is lovely, whatever is commendable…anything worthy of praise." Just like all media, games impact the attitudes of our hearts, and we need to teach our children to discern the themes promoted.

6. Does playing this game make your child more present?

We want our children to spend their time wisely, hanging their hope on Jesus, not escaping from the monotony, reality, and hardship that might arise during the day. It's fine to play a game because it's fun or beautiful or engaging, but it's something else entirely when we turn to games for satisfaction.

All art, games included, can help us worship God—we know if it's propelling us toward him when we produce good fruit. Love, joy, peace, patience, kindness, goodness, faithfulness, gentleness, and self-control pour out when we abide in Christ (Galatians 5:22–23). But if gaming results in the fruit of the flesh rather than the Spirit, then that's something to avoid. If it produces "sensuality, idolatry…enmity, strife, jealousy, fits of anger, [or] rivalries" then it's no longer building up and no longer a good fit (Galatians 5:19–21).

If you're in on games as a family, then you can use these six questions as a framework for conversation around which games to play. Games are awesome, but not perfect. There are video games that can draw us into wonder, relationship, and fun that reflect the joy and beauty of God and his creation, and there are games that fail to do these things. Our goal is to game *on* purpose—not *for* purpose.

Part 3: How then shall we live?

CHAPTER 7

Let's Talk About Porn (Resilience)

As someone who discovered porn in seventh grade, I can tell you first hand: Porn is terrifying.

My friend snuck some pictures into a card game box and brought it over so we could all sneak glances. Yes, I'm old enough that my first mistakes with porn were made with paper. My mistakes took time and planning.

But in today's tech world, and the blazing-hot speed and constant accessibility of the internet, porn is just one click away. The internet is in our pockets with smartphones, on our wrists with smart watches, and in our classrooms with one-to-one devices. The reason we must be equipped to engage this conversation is because porn does more than just amuse; it trains.

Fight the New Drug, an anti-porn organization, notes that pornography manipulates the way our brains work to make it difficult to walk away from. Pornography "can take our brains' natural stimuli…and give us more quantity, more exaggerated, and more 'supernormal' versions of that thing, until it can override what we think is normal, warp what we perceive as exciting, and make real intimacy seem less interesting by comparison." In other words, porn trains our brains to crave a type of interaction, with a type of individual, that doesn't exist in real life. Watching porn isn't just bad for us and others, but the very act of engaging or not engaging it forms a habit.

Let's pause here for a second.

Many (perhaps even most) of you want me to tell you I can porn proof your kid; that if you follow x, y, and z, your children will never open a nude pic (God forbid they *send* one), never stumble upon full frontal while scrolling Netflix, or wander into the dark waters of virtually any social media app (did you know you can access porn on Spotify? It's everywhere.) And while I wish I could concoct this magic elixir, we simply must acknowledge the reality of the digital world in which we live. To say we can porn proof our kids is to live in willful ignorance.

Children will come across their first pornographic image somewhere between the ages of eight and twelve. Some hear that and think "but not my child," and the sentiment is admirable. You've put up good boundaries, had the tough conversations, and made sure to be intentional with your tech. Yet there is a temptation to move beyond loving parenting into autocratic control—where mistakes can't be made because choices are never allowed.

Such helicopter parenting hopes to lock the worst of the world out with good enough rules and supervision, but doing so can actually make matters worse. Because the internet is wide-reaching in its accessibility, there is no group beyond the reaches of the porn industry; even the Amish struggle. As author Madeleine L'Engle writes, "Our responsibility to them [our children] is not to pretend that if we don't look, evil will go away, but to give them weapons against it." We don't fight against our children in this conversation—we must fight for them.

This is why we differentiate between porn *proof* and porn *resilient*: If our job as parents is to root our children in an understanding of the gospel, of God's design for his creation, and of his beautiful redemption plan, that must extend to conversations regarding our bodies, sex, and what media we choose to consume.

We will get to practical steps for making your family porn resilient, but before we do, I want to bring to the forefront three major problems with porn.

First, pornography changes the brain. It distorts a healthy understanding of sex and normalizes violent and aggressive sex. Second, it develops unhealthy and unsafe appetites through repetition. And finally, it creates a culture of abuse that preys on the weak and voiceless. Porn takes sex and makes it on demand; it takes something relational and makes it impersonal; and it takes something sacrificial and makes it consumable.

We fight for the innocence of our children—and against the monstrosity that is the porn industry–not by pretending like it's not there, but by teaching them to suit up for battle. We don't simply tell them "this is bad; avoid it," but train them to set their minds on the things of the Spirit.

We help them fasten the belt of truth, don the breastplate of righteousness (borrowed from someone much bigger than us), shod their feet with the gospel so they may be ready at all times to give an account, and protect their minds with the helmet of salvation.

The truths we share don't stem from parental positivity, but from God's Word that is living and active, sharper than any two-edged sword. So what do we say in raising them up in the way they should go? That we are fearfully and wonderfully made, that he knows the hairs on our heads (and the thoughts inside them), and the same Jesus who said our thoughts matter as much as our actions also said he is big enough to forgive them both. In doing this we don't simply keep them away from evil; we introduce them to the one who has overcome it and has called them to good works in a broken world.

With that in mind, here's what porn resilience looks like:

1. Remembering the gospel
2. Talking it out
3. Making a plan

Step one: Remembering the gospel

In my experience, most parents are scared about pornography and the impact it has had (or currently has) on their own brains, hearts, and relationships—not to mention their children's. This is why we need to remember where our hope is rooted.

Not even pornography is strong enough to keep your child (or you) from God's great love. Jesus has overcome the world—even porn. That doesn't mean we don't condemn what's evil, but we live in light of the truth that pornography, and all of its lies, entanglements and traps, are included and covered by the victory Christ won on the cross.

This is not an area hidden from God's sight. Rather, it is an area God can and does regularly win, cleanse, and redeem. Brains that have relied on porn to feel "normal" can be retrained, and the negative effects can largely be reversed. There is great hope in Jesus.

As Dr. Patricia Greenfield, director of UCLA's Greenfield lab for culture and human development, says, "A warm and communicative parent-child relationship is the most important factor [in reducing porn use among children]." We are all sinners in need of a savior, as Romans 3:23 says. Our goal as parents is to model repentance, not perfection. Through regular confession, we establish a baseline in the home of what it looks like to turn and trust in Jesus.

Step two: Talking it out

Our children are going to make mistakes. Sometimes even on purpose. While we'd all like to think that our kid won't be the one going and *searching* for porn, the truth is, they just might. Our job as parents is to make sure our kids know what to do when they do run into it (whether or not they intentionally went looking).

To this end, your goal is to have ten conversations about pornography by age ten. These conversations don't have to be the awkward and clammy conversations you may have had with your parents. Megan Flo of Birds & Bees, an online parenting resource on the subjects of sex and pornography, reminds us that it actually works best when they're just a part of regular life. "Bath time is a great opportunity to have those drip, drip, drip conversations with your little kids. When talking about body parts we want to be medical, respectful and matter of fact."

Sure, your kiddo may be silly about their body parts, and that is age appropriate. Flo urges parents to "stay calm and diffuse the silly and remove the shame" which, in turn, goes a long way toward modeling for our children that God made them, and their bodies, and that is good.

We live in the Pacific Northwest. Our many months of rain gives us beautiful rivers, massive evergreens, and robust undergrowth including, but not limited to, stinging nettles. Our boys love walking through the woods, spotting critters, and whacking plants with sticks. But few things ruin a tromp through the forest like a fistful of nettles.

So we teach our kids what nettles look like—every time we go out into the woods, we point them out. Our littlest is four and already knows how to spot the tell-tale heart-shaped leaf with a serrated edge.

But instruction doesn't end with prevention. Nettles have a way of sticking themselves into paths, just at the height where your ankle is exposed between your pant leg and sock. We also are sure to teach the medicinal properties of ferns and banana slugs, because most likely, someone will get stung—and we all need to know what to do when that happens.

The same is true with porn. We need to teach our children what it is, how to spot it, and how to avoid it. And we can also work to be sure that we all know what to do when things don't go as planned—how do we recover? How do we venture out on the internet again?

A good starting place is to define terms. What is porn? What is objectification? The book *Good Picture, Bad Picture* defines porn for young children as "people not wearing clothes being unkind with their bodies." Objectification is taking someone made in the image and likeness of God and lowering them to a means for our own satisfaction and selfish pleasure.

As our children age, we will need a more complex definition for porn because at its core, porn is anything we use to instigate or satisfy sexual lusts. Jesus makes it clear that the law indicts both those who participate in adultery as well as those who set their hearts on selfish gratification. In his words, "I say to you, everyone who looks at a woman with lustful intent has already committed adultery with her in his heart" (Matthew 5:28–29).

This isn't a lightening of the law, but a doubling down—

Jesus is warning us that we can commit unilateral adultery. A biblical understanding of porn absolutely involves blatant representations of sex, and Jesus extends that to show us it does not have to involve nudity. If we take the likeness of someone, whether they're clothed or not, whether they're real or a likeness, and sexually revel in it, it's porn.

Our goal here is ten conversations by age ten and to define our terminology.

Step three: Making a plan

Finally, we need a plan. When a neighbor brings over a laptop or your child searches "boobs" on YouTube, what happens next? Children need to know who to tell, what to say, and what to expect.

If we're operating in our homes under the banner of the gospel, then the good news is this: We are all sinners. Every day we will sin against God and against one another. Parents, this is an ongoing opportunity for us to model repentance. Being quick to own our mistakes and apologize for them creates a culture where it's safe to be human. And when sin is regularly confessed (not condoned, but not surprising), our children will know what to do with mistakes and struggles.

Who is the person your child will tell when they encounter porn? Have this conversation with them and be sure they know which trusted adult they can go to with this.

Next, give your child the words to say. I like the Mad Libs approach:

"Mom (or person you've agreed to tell), when I was on the computer downstairs (in a specific location), I saw naked bodies (whatever image made him or her uncomfortable)."

Tell your grown up. Say where you were (and who you were with, if applicable). And explain what you saw.

The goal is to begin a conversation, to hear what happened, and allow the parents to determine what next steps might be needed. Sometimes all that's needed is a celebration that your child confessed his or her mistake. Chris McKenna, digital safety expert and founder of Protect Young Eyes, encourages parents who receive such confessions to celebrate like they just got straight A's…"Instead of hiding it, they told you." We want to honor the confession; it's here, in the light, that freedom is found.

Building porn resilience in your home needs these three steps: Remember the gospel. Talk it out. Make a plan. It can be more, but it can't be any less. These last few chapters have been full, and I know, heavy. Digital spaces certainly have redeeming qualities, but when we're talking about them in regards to children, it can quickly become overwhelming. In the next section of the book, we are going to get very practical. I assure you, you'll end with hope rooted in the gospel and tangible tools to implement to make tech safe at home.

CHAPTER 8

Assessing Healthy Tech Use

After teaching middle school, and now speaking on tech use, I find myself returning to this central question: How can we know the difference between tech that is simply new-fangled and different from what I'm used to, and tech that is problematic and should be avoided?

Not everyone who uses Drool Tech uses it in a problematic way. I am certain there are video-gaming, show-and-music-streaming, social-media-posting, podcast-listening people who love Jesus and are healthy and well. Yet the question remains in my own mind: How much is permissible and how much is beneficial? The more I consider my own habits and tendencies, the more God is exposing my motives behind my tech use.

In his book *The Ruthless Elimination of Hurry*, author John Mark Comer points to our digital devices as culprits in helping us accomplish "more" while abiding in Christ less. "What you give your attention to is the person you become," he writes.

Paul put it like this: "For those who live according to the

flesh set their minds on the things of the flesh, but those who live according to the Spirit set their minds on the things of the Spirit" (Romans 8:5).

Where we set our minds, where we give our time, focus, and money, this is who we become. Jesus calls it abiding—the act of being present in the moment with God (John 15). Many times, we feel anxious, busy, or like the Lord has given us more than we can handle. And it's true; God will often give us more than we can handle to bring us to the end of ourselves and to him instead. But it's also possible that the constant accessibility and availability that tech offers hinders our ability to be present with others, let alone God himself.

So how do we know what healthy tech looks like in our own lives and in the lives of our children? This next section will walk you through what I call a RESET, a tool designed to help answer that very question. I recommend you begin with yourself. Children are expert hypocrisy spotters, so if you're not honest and open with your own tech (mis-)use when you come to them with this tool, it will likely fall on deaf ears.

The RESET looks at five areas and asks: Does this tech impede my:

- Relationships and Responsibilities
- Emotions
- Sleep
- Enjoyment
- Time

It's a simple yes or no answer for all five. No explanations, no justifications. If you find that's where the conversation is heading, it means it's likely an impediment.

Does this tech impede my:

Relationships and Responsibilities (YES/NO)

As Christians, we want our yes to be yes and our no to be no, even when it's inconvenient (Matthew 5:37). We are called to do everything unto the Lord (Colossians 3:23), and in doing so, we love God and others well in the responsibilities and relationships we've been given.

Does Drool Tech improve or impede relationships and responsibilities? Consider how it impacts family time, work, sports, school, homework, chores, community, and church involvement. Do you find yourself putting off commitments for a few more minutes of a game or app? Or reacting poorly when screen time is interrupted by a member of your household?

If the tech in question improves these areas, great! If red flags are going off as you read these questions, this is an opportunity for growth. It's for freedom that we've been set free (Galatians 5:1); we practice repentance, hand over areas of struggle, and set up parameters to establish healthy habits.

Emotions (YES/NO)

The easiest way to assess the emotional impact of tech is consider: How do you react when you don't get your tech time?

If you miss your show or gaming time or don't get to jump on social media, does it cause a spiral in disappointment or frustration? Do you get aggressive, snippy, or despondent when you don't get this time? Keep in mind—this reaction doesn't necessarily mean the tech itself is bad, but that it's taken up an unhealthy residence inside your heart.

When our oldest son was four, we were looking for something besides *Daniel Tiger* to mix into the show rotation. A friend recommended another PBS show, and we gave it a shot. However, after watching this new show a few times, Owen quickly became aggressive, argumentative, and hyperactive. These are fleshly responses that we were (and still are!) working to submit to the Spirit. While this show was fun and educational, it caused a change in behavior unhelpful in cultivating self-control and gentleness. Back to *Daniel Tiger* for us.

Sleep (YES/NO)

If there's Drool Tech in the bedrooms, sleep is being impacted.

How can you tell if sleep is being disrupted by tech? First, assess if there is Drool Tech in the bedroom. The blue light emitted from screens is enough to mess with melatonin production, the hormone that causes natural sleep to occur. Using a smartphone in bed, for example, increases the amount of time it takes to fall asleep. Second, if phones are next to the bed (or in the bed) and notifications are going off throughout the night, sleep is going to be interrupted and unproductive.

The fear of missing out paired with a false sense of urgency to respond to any and all notifications makes sleep fitful at best. If you're unsure, check the Screen Time assessment on the phone, and you'll see an hour-by-hour breakdown of app usage. In fact, research shows that the simple step of putting your phone out of your bedroom increases your happiness.

Here's why sleep is so important, especially for our children: It affects literally every part of development. According to the U.S. Department of Health and Human Services, a growing young adult needs nine hours of sleep a night. Sleep is crucial for developing minds.

According to Matthew Walker, professor of neuroscience

and psychology at UC Berkeley, "When sleep is abundant, minds flourish! And when it is deficient they don't." It's a sad downward spiral that affects intelligence, emotional intelligence, learning, memory, health, immune system—literally every bodily function.

In case that's not enough for you, a lack of sleep has been linked to an increased risk for depression and suicide. Too little causes our bodies to go into fight-or-flight mode, resulting in what we commonly call panic attacks. Our children live in a world full of potential distractions and stressors—bedtime should not be either of those.

Enjoyment (YES/NO)

When we talk about how Drool Tech impacts enjoyment, what we're asking is this: Does it allow you to enjoy and flourish in the special ways God has created you?

An easy assessment is to look at the fruit your tech time produces. After using Drool Tech, are you more filled with love, joy, peace, patience, kindness, goodness, faithfulness, gentleness, and self-control? Or is it producing attributes more in line with lust, hate, strife, jealousy, anger, and rivalries (Galatians 5)?

Our actions will flow from the condition of our hearts. If we are using tech from a place of security and hope in Jesus, then our posts, music stream, search history, games, and shows will reflect the joy we have in him. If, instead, we use tech to strive after purpose apart from Christ, then that will show as well.

The goal is to enjoy all the tech that helps us love God and others, living out the works he's set before us to do, and none of the tech that distracts us from this.

Time (YES/NO)

Finally, the last self-check on the RESET is to consider how our tech use impacts our time. One area to keep watch for regarding time is how well we (or our children) maintain time parameters. Can we stick to the agreed upon limit? Or are we consistently scrolling for a few more minutes, playing one more level, watching one more show? Are we sneaking time with tech?

While the amount of time spent on screens does matter (the general rule is less than two hours a day on screens), the heart behind how we spend our time matters too. For Christians, we must "number our days, that we may get a heart of wisdom" (Psalm 90:12). Our standard for how we spend our time moves from "not that bad" and onto "whatever is best"—that is, whatever helps us love God and love others. There are nearly infinite ways we can spend our time, skills, and efforts, but just because we can doesn't mean we should.

A RESET helps us assess if tech is healthy or not. The backbone of it, though, goes deeper than determining if something has a screen or plugs into a wall; rather we acknowledge that we are all sinners in need of a Savior, and we all tend to take good things and make them into little gods. Our goal, then, isn't to get rid of tech, but to see our children become more like Jesus after using it.

CHAPTER 9

Build a Hedge

As we enter into the specifics of making tech safe at home (and I promise, we will get very practical in this chapter), I want us to remember the bigger picture. We're talking about tech at all because of two things:

1. Our hope is rooted in Christ alone—not our ability to be perfect ourselves or parent our children perfectly and avoid all tech pitfalls.
2. We are aiming to raise children who love God and use tech. In that order.

With that as our baseline we can talk about making our homes safe. The best way to have healthy tech at home is to have intentional boundaries around our network, devices, and family. Healthy boundaries make for healthy tech.

I love the image of a hedge. We want to put distance between our family and tech; we're not building impermeable walls, or a bunker, just barriers that naturally give us space to grow, talk, and cultivate character and relationships. In this way, we can choose to engage tech on our own terms and timing.

To some, the idea of hedges screams behavior modification. It feels like this idea of using tech from the gospel should be entirely internally driven, and any external limits or restraints are simply religious makeup on an otherwise rebellious pig. Some parents argue that, in the case of true-believing children, God will keep their hearts from making the worst mistakes because they love him. From this perspective applying restraint to digital tech is a sign of lacking faith in God's good work in our children's hearts.

But I don't tell my kids to put on seat belts to test their obedience or gauge God's faithful providence in numbering our days. I don't drive a car with airbags out of fear or because I don't trust the Lord. Instead these devices mitigate some of the dangers of driving the open road in a broken world. If I acted out of fear, I'd avoid driving all together and such a decision would lead to a host of similarly logical conclusions I would also have to cut out due to the intrinsic dangers involved: stores because of crowds, parks because of strangers, and church because of sinners.

Instead I believe the risk of driving is worth the benefit, and I believe my children are worth protecting. Seatbelts and airbags are blessings designed to help my children be safe despite potential dangers. It's the same with tech.

I don't agonize over all that could go wrong, but plan for how to maximize what is going right. I've already purchased the "vehicle of the digital age" (the computer, smartphone, smart TV, or tablet that drives on the internet highways); now I need to be sure I don't skimp on seatbelts and airbags. In this way, setting up boundaries isn't stopping all mistakes but making good decisions easier.

The best way to begin is to build a hedge. In fact, we're going to build three. Imagine three concentric circles, gradually gaining height as they move away from the center. This is our goal for protection.

First, we build a hedge around our home internet.

If your home has internet, you have a network. This is the large umbrella of protection that covers every device that doesn't have internet on its own (tablets, laptops, desktops, e-readers, old smartphones, gaming systems, smart TVs).

If you have wireless internet at home—not the cell service from your smartphone—you have a device called a router. Somewhere in your home there's a cable running from the wall that plugs into a box or tower, and that's the box we're talking about. The router provides you with internet, and it's your best friend when it comes to building a hedge around your home internet.

A strong router ensures that everyone who uses your internet has to play by your family's tech rules.

There are two options:

Buy a new router. Routers are that little box that sends wireless internet to your entire home. If you have internet on a device that doesn't have a cellular plan (i.e., your smart tv), then you've got a router. Certain routers are designed to enable parents to easily set time limits, create search-term rules, and apply user protections for every device using the network. If you're in the market for a new router, just want faster internet, or are looking to improve the protection of your home network (keeping prying eyes off baby cameras, for example), then a new router is a great option.

Add protection to your existing router. For those of you who are happy with your router, either because your current one does enough or you've recently upgraded, then adding protection to your current router is a great option. These after-market additions will allow you to set time and content restrictions as well as provide network supervision.

Whichever option you choose, you gain the peace of mind knowing you have set up loving and intentional protections around the number one way your family is going to engage the internet: Your home internet.

The goal isn't to build a wall so high that no one can make a mistake, but to protect against simple mistakes and carnal curiosity. This is also the only way to truly manage devices that cannot have programs loaded onto them like smart TVs and school-issued computers.

Second, build a hedge around your devices.

Not all devices rely on your home internet. Smartphones, watches, cell-equipped tablets, and such need support of their own. These devices need apps installed on them to stay accountable even when they're away from home. This layer of protection is a great way to keep the healthy tech conversations going even when everyone isn't in the same room.

At the risk of dating this chapter quickly, I'm including some resources that will equip you to do this. The most up-to-date resource list is at gospeltech.net/parentresources. Or you can search "resources like ________" and insert one of these options. As of writing, these are the market leaders in the effort to help families default to safety online:

1. **Bark:** This is, at the time of writing, the best option for monitoring social media. Bark specializes in looking for concerning phrases, emoji combinations, and sends you a "bark" notification if it finds something. It's a great option for allowing your child the trust he or she earned online while still verifying that they are safe and healthy.

2. **Covenant Eyes:** This is a great resource for accountability with pornography. Covenant Eyes scans everything that comes across the device's screen: texts, shows, internet searches, etc., and notifies a pre-selected ally of concerning content. Their target demographic is ages sixteen plus.
3. **Canopy:** This blocks pornographic content in real time. Canopy is an excellent option for devices accessed by children and young adults; but as children age, we will need to scaffold the use of blocking devices. Our goals for online safety will look different for a six-year-old than a sixteen-year-old. We aren't trying to forever shield them from the dangers of the internet, but instead train them to make their own wise decisions online.

All three of these options work best on personal computers, desktops (Apple or PC), and Android devices (Samsung or Google products). The iPhone prioritizes privacy over safety, which means that reporting apps like Bark and Covenant Eyes cannot see what happens within specific apps. While the iPhone can be locked into a safer mode (using the Covenant Eyes app for a browser and not allowing app downloads, for example), anyone looking to truly make smartphones safer will be better suited with an Android device.

Finally, we build a hedge around our family.

This last step is intended to be a conversation with all members of your family who are using tech independently, as the goal here is to develop a Family Tech Framework. Any child old enough to be using tech can helpfully participate in developing guardrails for family use.

Below, you'll find the six pillars for developing a Family

Tech Framework. As any classroom teacher will tell you, establishing norms guide expectations—if we don't develop and communicate expectations, we set ourselves up for frustration and failure. This is what the Family Tech Framework (FTF) is designed to do: Empower parents to talk about healthy tech and create family expectations with their own children.

For a more thorough example of Family Tech Framework, see the Appendix. For now, here is a simple overview of the areas of the Family Tech Framework you can develop.

1. Set ground rules.

Effective tech ground rules set expectations that are clear, biblical, and agreed on by the whole family. Similar to when we play a game of pick-up ball, it goes better when we mark the end zone clearly or determine we're playing to fifteen. Ground rules mean we're all operating with the same basic understanding of healthy tech. One of our favorite ground rules is, "All expectations in the framework apply to all family members and any guests in our home."

2. Determine tech time.

Every family member needs to know how much time on tech is permissible, when that time can be used, and how often. There's a huge difference between using tech whenever you want, however long you want, versus using tech with self-control and restraint. We want to do more of what Andy Crouch encourages in *The Tech-Wise Family*: Keep tech in its proper place while it waits for us.

For example, maybe Drool Tech can be used during the hour before dinner. Determining tech time helps form rhythms, keeps people present for relationships, and trains us away from compulsive tech habits.

3. Establish tech locations.

Now that we know *when* we can use tech, we need to specify *where* we can use it. Are laptops allowed in bedrooms? What about tablets or smartphones? TV? Gaming systems? Any and all Drool Tech that is under your roof needs to have specific locations for use—the goal here is to set all members of the family up for success, because often, it's when we're isolated and think no one's watching that we give into temptation.

This also provides a chance to discuss expectations for using mobile devices outside the home (on the bus, at school, in the car, at friend's or neighbor's houses.)

4. Assess content.

As a family, determine what content is acceptable and unacceptable—and why. Paul instructs us to set our minds on the things of the Spirit, and warns that "to set the mind on the flesh is death" (Romans 8:5). Setting our minds on the things of the Spirit is an intentional act and takes practice, especially when we've become lax with our Drool Tech.

Our family's goal is to submit all content to what we call the Philippians 4 test: "whatever is true, whatever is honorable, whatever is just, whatever is pure, whatever is lovely, whatever is commendable, if there is any excellence, if there is anything worthy of praise, think about these things."

5. Outline priorities.

We either set our priorities, or someone sets them for us. Without a clear vision of where we're going, we're bound to let lesser things become primary things. Instead, we want to be intentional about our family's vision and purpose, walking in the path God has set before us.

We begin by determining what things come first with our time, calendar, money, and energy. Everything else fits around. This is what, when my sons ask when they'll get to play video games, I can easily show them how we spend our days and weeks, making it clear that something else (like community group, time to play board games, see friends, read, play sports, or practice piano) would have to get bumped; they can see for themselves that, at this stage of their lives, video games just don't have a place.

6. Prioritize emotional and physical safety.

Equipping children to make wise and safe decisions online begins at home. While we want them to be innocent as doves, they also need to be able to walk wisely in the world. Children need to know the potential dangers in the digital spaces they're occupying.

A child who is watching *Daniel Tiger* on a locked iPad needs a different safety conversation than a child scrolling TikTok independently. Topics you might cover here, depending on the age and maturity of your child, include: strangers, bullies, porn, as well as the mental effects of time spent online.

This, like many of the other guardrails, is a conversation we will need to have repeatedly. Tech is constantly evolving and our children are constantly changing and growing too. The first round of your Family Tech Framework will not likely see your children from six to eighteen. It will need to be revisited and perhaps modified to adapt to the ages and stages of development.

Making tech safe at home is a mix of intentional conversations rooted in relationships. We certainly can make practical choices with the hardware and software we choose, but ultimately, the goal here is to guide you in an on-going conversation, rooted in the hope of the gospel. We guide our children, not simply away from the bad, but toward the very best things God has for them. We want them to use all the tech that helps them love God well, and avoid any that does not.

CHAPTER 10

Raising Healthy Youth in a Tech World

How do we know if and when we should adopt new technology into our homes? When your child, the light of your eyes, walks into the room like a modern Veruca Salt and says, "Daddy, I want a pony," what will you say? More importantly, *why* do you say it? Raising our children in a tech world means more than just denying them "bad" tech; it means training them to think about the tech they use. We want young people who are as wise as serpents, able to discern the benefits and drawbacks of the tech they use, and as innocent in the mistakes and ways of the world as a dove. We want them to use all of the tech that will make them more of who God has designed them to be.

Adopting New Tech

First, consider the work and effort that goes into acquiring a new pet. Generally, before agreeing to the responsibilities of a new animal, we look into the cost, work load, engagement level, and the physical space we occupy; we might also consider how responsible our child is and the likelihood he or she will actually help care for Fluffy.

This is why some of us end up with Great Danes and others of us end up with Beta fish. It's why, on more occasions than I'd like to admit, we have offered our beloved family dog to family and friends who have more capacity for caring for a high energy pup than we do. You see, when we got Ruby, we were in a season of wanting companionship; we were also living that no-kid-dual-income life, with lots of free time and disposable income.

Today, we are the opposite of that DINK life—we are living that oft-coveted "one-income-three-kid" life. Our capacity and threshold for adding animals to our home is zilch. Despite our children's bartering and begging for any and all animals they walk past at PetSmart, the answer is always, "When you're a grownup, you can have all the snakes/gerbils/birds/cats/tarantulas/dogs you want." In other words, *no*.

If we put such thought and care into deciding upon what animals to have in our home (when, largely, the research supporting benefits of caring for a pet is ubiquitously positive), we should be doing the same before adopting new technology into our homes (where the research is fraught with contradictions at best, and proves detrimental at worst).

Before we acquire new technology—or give into a particularly persuasive child—I want to challenge you to consider three questions:

1. Is it safe?
2. Does it line up with family expectations?
3. Does it cultivate your child's potential?

As loving parents, we want to give our children good gifts. As tired parents, we want five minutes of peace and quiet.

But before we hand over tech, let us revisit our starting point: Our goal is to raise kids who *love God and use tech*. When we are considering adding tech to our home, we must keep the endgame in mind and ask ourselves how this tech will help achieve the goals we've determined as a family.

Let's consider first: **Is it safe?**

When we're determining the safety of tech, we are looking at four areas:

1. Is it Tool or Drool Tech?

The type of tech we use matters. Drool Tech, remember, wants to take your time, focus, and money. Its goal is for users to stay engaged because that's good for business. But we want to buy and use tech that supports our own goals; using a lathe for carving bowls or a 3D printer for designing robots is drastically different from playing video games or scrolling social media.

2. Does it have a web browser?

Web browsers come on an array of devices and are the portal to the amazing—and sometimes dangerous—world of the internet. While safety settings can help, few devices default to anything like parental mode. Before adopting a new technology, assess your comfort level with creating safety around your device and network, and determine if a browser on this device is a good fit for your child, family, and season of life.

3. Does it have apps?

The apps that keep us coming back for more—and the ones that keep us longer than we ever planned to stay—are designed as Drool Tech. No one (okay, maybe *someone*) ever stayed too long on the weather app or spent too much time using the calculator app.

The trick here is that apps offer suggestions for content you weren't looking for in the first place: This is how YouTube gets you to stay, how Netflix keeps you watching one more episode, and how Instagram keeps you browsing through photos of strangers. They are *designed* to do just this. Before adding new tech, consider what apps are available on it and if they're worth your child having.

4. Can it be held accountable?

My friend Ryan Frederick, founder of Fierce Marriage, coined the term "the lie of anonymity" when discussing how we are prone to compromise our integrity more easily when online. The internet promises (and fails) to keep your secrets and let you be whomever, and whatever, you want to be.

Accountability makes sure we live in the light—and above reproach. If you haven't already talked with your child about how the internet never forgets, add it to the list of conversations to keep circling back to. If the device you're considering doesn't allow for accountability to be added to it, perhaps consider if the risks are worth the potential cost to your child.

Let's consider if the tech in question lines up with our family expectations.

Raising kids in a tech-saturated world requires a clear standard for what is right and wrong; and while we live in a world that loudly says to follow your heart, find your own truth, or that there, in fact, is no absolute truth, the Bible quite clearly says the opposite.

In fact, it says that the heart is deceitful above all else (Jeremiah 17:9), that Jesus is the way, the truth, and the life

(John 14:6), and that the word of God is our standard for how we live (2 Timothy 3:16). As we've discussed previously in regard to crafting a Family Tech Framework, Philippians 4:8 provides us with a set of guardrails regarding our content consumption: "Finally, brothers, whatever is true, whatever is honorable, whatever is just, whatever is pure, whatever is lovely, whatever is commendable, if there is any excellence, if there is anything worthy of praise, think about these things."

When this is our standard for content, we are forced to move from a checklist of "good tech" versus "bad tech," and instead consider if what we're ingesting builds us up into who Christ is calling us to be. It's not as much about *how* we're getting our music (Spotify, Apple, CDs, or vinyl records), but if the music is true, honorable, just, and pure.

Now, with the content question in mind, look at the second half of verse 8 above and ask, *Is it lovely, commendable, excellent, worthy of praise?*

When I give this talk to live audiences, I can feel the air leave the room; this standard feels like a killjoy. What can we play, watch, listen to, and read if we take this verse literally? But let's remember, Paul isn't trying to fix our hearts with "better" content. He's giving us a way to see if what we're celebrating is helping us set our minds on the things of Christ—because setting our minds on the things of Christ is life, whereas setting them on the things of our flesh is death.

When Paul tells us in Romans 8 to "set our minds on the things of the Spirit," he's telling us two beautiful truths:

1. There is a tangible difference between the things of God and the things of the flesh.

2. We have a choice as to where we fix our minds.

Family expectations are to be designed with these two truths in mind. Our digital choices, even for our entertainment, matter to God because as Christians, we belong to him. Before adding any new tech to your home, keep this in consideration.

Finally, when deciding on new tech, examine how it supports your child's potential.

We've determined if the tech is safe and how it aligns with your family expectations, essentially determining that it's permissible. But is it beneficial?

When tech is beneficial, it will do two things: It will meet a need *and* it will develop your child's potential in Christ.

When we look at the function of something like a Gabb Phone, we can see it's a beautiful piece of technology. It's one I regularly recommend to families. And yet—I haven't given one to my ten-year-old. Why? While it's safe and supports my family's expectations, it doesn't fit a need or desire for my son. He has no one to call, is rarely without a parent or another trusted adult, and has never asked for one. Giving him a Gabb Phone in this season would serve as a distraction rather than provision.

Additionally, great tech will make your child look more like Jesus. When I said this recently at a live talk, it was met with peals of laughter. We don't typically think of tech as being something—especially regarding our children—that cultivates the fruit of the Spirit. It's typically quite the opposite.

There will come a day when our son will be in need of a phone like the Gabb—when he's more independent, involved at school, and has more freedom with his friends. As much as I'd like to teach him the ways of yesteryear (like calling Collect and yelling into the payphone, "Come get us from the mall!" when you're prompted for your name), this just isn't an option in today's world. Something like the Gabb phone will allow him to meet a specific need, while also allowing him to continue growing into the young student-athlete/friend/sibling God has called him to be.

Please remember that our standard for adding tech to our homes isn't whether we like it a lot, whether we've used it for years, or if it's won a lot of awards. We have a better standard, and we're called to steward tech in light of that standard. Just like all those cute animals at the pet store, tech requires mindful consideration, intentional conversations, and carefully counted costs before bringing anything home.

Building Tech Trust

Trust is a key ingredient of any relationship, and it's built with tech the same way it's built anywhere else: over time and with a series of small choices. It can also be lost rather quickly. Tech trust is central to our conversation about loving God and using tech because there are few areas that expose our hearts more quickly than our tech use.

Part of the reason for this is that tech makes some subtle, powerful promises. Most of our poor decisions regarding tech stem from us pursuing false hope.

Tech promises:

- Fame without accomplishment
- Influence without investment
- Knowledge without limit
- Intimacy without cost
- Belonging without being known
- Money without effort
- Adventure without risk

At the core of every one of these promises is a familiar call to personal agency. By that standard, we begin to live our lives on our own terms, and in doing so, make ourselves god. If we think back to my tech testimony, it's clear that I didn't stay up gaming all night because I thought it was good for me. I decided God was fine, but I *needed* games to be okay with daily life.

It's important to remember that a gospel-centered understanding of trust is not about never making a mistake, but about confessing quickly when we do. In this way, we may not raise perfect kids, but Lord willing, humble, honest ones who have seen what it looks like to repent well and often.

Why We Trust our Kids—*And* We Verify Their Actions

All these little steps we've walked through in this book are intended to help build trust, and also verify that your trust is well founded. Trust without verification is just wishful thinking.

Think in terms of a bank account: I might believe there is enough money to make a purchase. I might even trust it's enough to shop for a few items, but no amount of trust or positive thinking will add a penny to my bottom line. The transaction will either clear or be denied, and whichever outcome, my positive attitude towards my cash resources will have zero impact on the reality of it.

Similarly, I can't simply hope my children will stay safe online. I can't hand my child a smartphone, or any other internet-capable device, and wish for the Holy Spirit to stop their penchant for mischief while also holding off any intentional bad actors. Maybe my kid will be fine and avoid the pitfalls of the interwebs, but my positive attitude has nothing to do with it. Trust without verification is a set up for a relational overdraft. We want to make sure we're trusting within the limits of what's been earned and can be verified.

Finally—and this is key when it comes to trust—as loving parents, we cannot expect our children to self-report. Yes, communication is foundational for trust. Yes, we tell our children they can come to us with anything, and there's nothing they can do to make us not love them. That's true and biblical. But confessing is hard; it's a whole lot easier when someone presents the evidence to you in love and asks what's going on.

Does this excuse your child's active running and hiding? No. But remember that God didn't wait for Adam to confess his sins; he sought him out and kick-started the conversation with a rhetorical, "Where are you?" God didn't pretend Adam didn't sin, and he didn't shrink from the conversation either.

If you feel the need to know everything your child does with technology every waking moment of their life, please hear me: You don't trust him or her. Don't give that child a smartphone. But if they've earned your trust, and they've proven they can make wise, intentional choices in a digital world, then trust them. Don't spy on them.

Verification of tech use (and your child's trustworthiness) isn't determined by how "good" of a kid your child is. Driving a car safely isn't the mark of a good or bad person, but a person ready to do so. To be trusted with high-risk activities, like personal online freedom, one must be physically, emotionally, and spiritually prepared.

What might verification look like? We can use tech to support tech. As discussed, build a hedge around your devices, networks, and home—that is the foundation.

Please note: There's a huge difference between accountability and spying. Accountability is built on the trust your child has earned by being faithful in many other areas of life. You trust your child, you give them appropriate freedoms and have a consistent, open relationship in both real life and digital spaces. Accountability simply verifies that all is as it should be. When something goes wrong you immediately initiate a conversation, not punishment. If, however, you don't trust your child and feel they need constant, 24/7 surveillance, there's a good chance they aren't yet ready for digital, internet-connected freedom.

The foundation of tech trust is training our children to live *from* hope, not *for* hope, in their tech use. We don't simply want them to do the right thing; we want them to do it for the right reason. When our children understand the hope, freedom, and purpose from being made new in Christ, it will show up in their tech. When they make right choices online it won't be because they fear reprisal from their parents, but because they have great joy in Jesus. They will use their tech to extend the hope they have in Christ. That is my goal for every one of us, and especially so for the young people we are raising in this tech world.

CONCLUSION

Love God. Use Tech.

All of our work has come to this point—we've reached the summit of our conversation and we can look back at everything we've learned and put it all into perspective.

The research I've referenced throughout this work is momentous—the minds and hours that have gone into equipping us to parent well in a tech world is incredible—yet those facts alone don't have the power to help us apply them.

The apps, devices, and programs with teams dedicated to scaffolding safe tech use is encouraging and will change the way generations of people use tech; but we'll find new ways to make mistakes if there isn't more to it than clever programming.

My hope for this book is that you will love God and use tech, not the other way around. My hope is that you are equipped with the tools with which to raise your child in the fear and admonition of the Lord, and to do so in a tech world. You are aptly equipped at this point. You know the two types of tech, can spot if it's healthy or not with a RESET, and have the basics for building a hedge around your network, devices, and family. That's all the tech knowledge any person needs to get started.

There's one more thing: The best hedges in the world can't protect our children from a heart that wants to make a mistake. So what do we do when our prayers are in line with St. Augustine as we cry out, "Lord, make me good, but not yet"? What hope is there for us when we make the wrong choice, and we do so on purpose? Much, in every way. We who are new creations cry out with the apostle Paul "O wretched man that I am! Who shall deliver me from this body of death? Thanks be to God through Jesus Christ our Lord" (Romans 7:25–26)!

This is the point. When we use our tech from the love God gives us, we can actually use it well. Yes, we'll have boundaries, but those are in place as reminders of the love we already have. It's not because a mistake might lose us anything of real value, but rather that we already have the real value and we don't want to be distracted pursuing anything less than what we already have. If we seek out any sin as a Christian, and say "This will satisfy me," we lie. We know the only one who can satisfy our hearts and souls, and all our rules and accountability are meant to point us back to that one great Truth. This is the hope of the gospel as it applies to our tech.

The same is true with our standards for entertainment, creative endeavors, and engagement through technology. We no longer hold ourselves to a comparison with others ensuring we're "not as bad" as those around us. Instead we reflect on what God has called us to—that everything we set our minds on is true, honorable, right, pure, lovely, good, and excellent (Philippians 4:8). This gospel focus reminds us how we can do any of this good work in the tech space—we love because we were first loved (1 John 4:19).

So let's go and do that. Let's love God with all our hearts, souls, minds, and strength, and love our neighbors as ourselves. Let's do it in person and let's do it online. Let's use all of the tech that makes us look more like Jesus, and skip the rest. Let's put the gospel at the center of our tech use and our families so that we can go out and love God and use tech.

Appendix

From Chapter 4: Your Kid Doesn't Need a Smartphone

Tools for healthful discussions of pornography:

- *Good Pictures, Bad Pictures* by Kristen A. Jenson & Gail Poyner (also available in *Good Pictures, Bad Pictures, Jr.* for children ages 3–6)
- *God Made All of Me*, by Justin S. Holcomb & Lindsey A. Holcomb (for ages 3–6)
- Fight the New Drug (fightthenewdrug.org)
- Dumb Phones (available at time of publication)
- Gabb Phone: Tool tech to the max. This is an excellent starter to begin building trust. Designed for when your child needs to communicate for friends and sports, and no distractions. (https://gabb.com/)
- Bark Phone: Defaults to safety. Bark does an amazing job scaling with a solid parenting app that allows you to ensure you'll be the first to respond when mistakes happen online, and also allows your child the digital

trust they have earned by being faithful in the little things. (https://www.bark.us/bark-phone/)

- The Wise Phone II: This phone is built for anyone who wants less distraction and more tools for getting the most out of life. The Wise Phone II is built around intentionality. It can make calls, text, use GPS, and take high-quality photos, but it doesn't have an app store or web browser. It's the perfect option for youth and adults who want Tool Tech without the distraction. (https://techless.com/products/wisephone-ii)

From Chapter 5: Social Media is Bad for Your Child's Health

These books speak to tech use, screen addiction, and social media; all have shaped how my family and I think about and use technology in our home:

1. *The Tech-Wise Family: Everyday Steps for Putting Technology in Its Proper Place*, by Andy Crouch. Crouch encourages each of us to ask why we have the tech we have, and what it adds to our lives. His goal is for all of us to keep tech in its proper place. This book particularly shaped how we think about tech in our home.
2. *Alone Together*, by Sherry Turkle, MIT professor. Turkle argues that we demand more from tech than we do from one another—partly because tech promises us relationships without cost or risk.
3. *American Girls: Social Media and the Secret Lives of American Teenagers*, by Nancy Jo Sales. Sales raises the alarm that social media is not simply connecting

kids with kids, but throwing them into the wider world of internet content. Through modern media, she argues, girls are sexualized and trained to be so, which worryingly spills over into the real world through what is expected, accepted, and even praised in social media and entertainment.

4. *Glow Kids: How Screen Addiction Is Hijacking Our Kids,* by Nicholas Kardaras. Kardaras wrote *Glow Kids* in 2016, which felt almost too soon to be hitting the nail on the head about the impact of technology on developing brains. A reading now reminds us how far our conversation about tech has come—we understand screens aren't in fact the problem; the content is—and also how much we still have to learn.

And these books have shaped how we think about tech in light of a bigger conversation around the trends and patterns in entertainment, parenting, and the church:

5. *Amusing Ourselves to Death: Public Discourse in the Age of Show Business* by Neil Postman. Postman is clear from the introduction to his novel that his fear wasn't government crackdown on liberties but rather societal pursuit of pleasure at all costs.
6. *The Coddling of the American Mind: How Good Intentions and Bad Ideas Are Setting Up a Generation for Failure,* by Greg Lukianoff and Jonathon Haidt. These authors draw attention to three great untruths: The great untruth of fragility: What doesn't kill you makes you weaker; The great untruth of emotional reasoning: Always trust your feelings; The great untruth of us versus them: Life is a battle between good people and evil people.

As I mentioned in the chapter, the Family Tech Framework is a resource intended to empower families to talk about healthy tech and establish tech expectations in six areas: Ground rules, time, place, content, priorities, and safety. (https://gospeltech-workshop.com)

From Chapter 6: Video Games

See "Video Game Assessment" tool on page 122.

From Chapter 7: Let's Talk about Porn (Resilience)

For resources on equipping and encouraging parents and care-givers in the area of porn resilience, check out:

- *The Gospel Tech Podcast* Episodes 11, 126, and 159
- Birds & Bees (@birds__bees on Instagram)
- Greta Eskridge (@maandpamodern on Instagram)
- Fight the New Drug (fightthenewdrug.com, @fightthenewdrug on Instagram)
- Proven Ministries
- Protect Young Eyes (protectyoungeyes.com, @protectyoungeyes on Instagram)
- Covenant Eyes: Victory App (free)
- Samson Society (samsonsociety.com)

From Chapter 8: Assessing Healthy Tech Use

See the RESET Assessment on page 126

From Chapter 9: Build a Hedge

Building a hedge around your family's tech use means investing. Here's what I recommend:

For your network:

- Get a great router. The Gryphon Router is what we chose for our home (we use the AX model). It's super fast and allows me to set time limits, search regulations, and create user protections for every device that logs into my network. (https://gryphonconnect.com/)
- Power up your current router:
    - If you like your current router, you can add powerful protective features using something like the Bark Home. This will allow you to set time and content restrictions as well as network supervision. (https://www.bark.us/learn/bark-home/)
    - The Circle is another solid option for adding parental controls to an existing home network. Just plug the device into your router and you will have the ability to set time limits, monitor online activity, allow and block specific apps, and start those important conversations about how to use tech well. (https://meetcircle.com)

For all your devices:

- Smartphones should be accountable for family members of all ages. I strongly recommend Bark for young people who have earned the right to be online and on social media. (https://www.bark.us)

- One of the best things we can do with our personal devices is remove the lie of anonymity—that subtle lie that VPNs and incognito mode tell us—that we are the only people who have to know what happens online. The truth is God sees everything, and our tech should reflect that reality. I personally use Covenant Eyes for such accountability. While it has filtering options, the real strength is ensuring that whatever you see online someone else who cares about you will also see. This focuses on loving relational accountability rather than fear-based internet surveillance. (https://covenanteyes.com)
- Canopy provides smart filtering that can block images in real time within browsers as well as websites. It also protects devices from sending and receiving inappropriate images and also custom URL, app and category blocking. Canopy is a great option for devices that younger family members use. (https://canopy.us)

For your family:

- As I mentioned in this chapter, the Family Tech Framework is a resource intended to empower families to talk about healthy tech and establish tech expectations in six areas: Ground rules, time, place, content, priorities, and safety. I've created an online course, and a PDF workbook, for families and small groups interested in getting more support in creating and implementing a Family Tech Framework (https://gospeltechworkshop.com).

Video Game Assessment

Should My Child Play This Game?

You can answer this question in three simple steps:

1. **Be gracious.** A game is a thumbs-up until proven otherwise. A game earns a thumbs-down each time it doesn't meet your family values and expectations.
2. **Be humble.** Always have these grades be part of a conversation, not the final answer.
3. **Be Informed.** Use these resources to answer the questions in each category:

a. Check with the pros at ESRB (esrb.org).

b. Read what parents say on Common Sense Media (commonsensemedia.org).

c. Discover average time to beat at How Long To Beat (howlongtobeat.com).

d. Observe actual gameplay on Youtube or Twitch.

e. Search online content for any additional information related to the game using Bing or Google.

With the above groundwork done, you're ready to get started on the next page.

Six Questions to Grade a Game

Category	Question	Yes?	No
Characters	Would I have this character over for dinner?	+1	0
Themes	Do the lessons taught line up with my family's standards?	+1	0
Violence	Is violence part of the game or the point of it? (Think: Mario Brothers vs. Mortal Kombat.)	+1	0
Language	Is the language in the game helpful or hateful?	+1	0
Play Time	Can this game be beaten?	+1	0
Game Experience	Does this game make my child more present? i.e. Does it produce good fruit? (Galatians 5:22)	+1	0

Game Score: ___ / 6 points

Score Breakdown

6 points: Play it! Excellent match for your family.

5 points: Talk it out. Mostly ok, but off in one important area. Is it worth it?

4 points or less: Skip it! This game varies in two important areas from your family standards. Some fun just isn't worth it. Talk about alternative games or activities.

(continued on the next page)

Reflect and Apply

Now you have a grade; use it to discuss whether this game is a good fit for your family.

Some verses to keep in mind. Our games should:

- Celebrate whatever is true, honorable, just, pure, lovely, commendable, excellent, and praiseworthy. (Philippians 4:8)
- Avoid celebrating a "debased mind:" "unrighteousness, evil, covetousness, malice. They are full of envy, murder, strife, deceit, maliciousness. They are gossips, slanderers, haters of God, insolent, haughty, boastful, inventors of evil, disobedient to parents, foolish, faithless, heartless, ruthless." (Romans 1:28–30)
- Strive to be more than just "not that bad." Just because it's not harmful doesn't mean it's beneficial. "'All things are lawful," but not all things are helpful. "All things are lawful," but not all things build up." (1 Corinthians 10:23)
- Produce the good fruit of love, joy, peace, patience, kindness, goodness, faithfulness, gentleness, and self control. (Galatians 5:22)

In a phrase, our games should make us look more like Jesus.

The RESET Assessment

Time to take the Gospel Tech RESET Assessment. Read each question, and then answer each question "yes" or "no". Each "yes" will be added into your RESET score at the bottom of the sheet and shows an area where tech is distracting your purpose.

Each family member should assess their RESET separately, then come together to discuss. The aim of this conversation is to make your tech use line up with your call and potential.

Pro Tip: God tells us through James that we ought to be quick to listen, slow to speak, and slow to become angry (James 1:19). Prayerfully prepare yourself to give loving feedback to others and to hear constructive feedback on your own RESET assessment.

Add your "YES" answers to find your RESET score:

___ / 5

score of 0 = healthy RESET
score of 1-5 = areas to improve

DOES DROOL TECH IMPEDE MY...	NO	YES
R: RELATIONSHIPS + RESPONSIBILITES?	I am fully engaged with healthy relationships. My work, sports, school, homework, chores, community involvement and family time are not impeded by tech use.	At least one relationship or responsibility is not at full capacity because of my Drool Tech use.
E: ENJOYMENT OF DAILY LIFE?	I enjoy life to the fullest in all the ways God has uniquely made me.	Sometimes Drool Tech is just more fun than the stuff I used to enjoy. I might still enjoy other stuff, but I just don't do it as much.
S: SLEEP AND REST?	I keep all Drool Tech, including my phone and personal entertainment systems, out of my room. I get a full night's rest and interruptions don't come from tech.	Tech is the last thing I see before bed and the first thing I see when I wake up. I have assorted Drool Tech in my room. My phone is my alarm.
E: MY EMOTIONS?	Life is hard and I've learned how to process emotions in relationship. I can bring up hard questions, listen to others, and I'm learning how to process in a respectful and loving way (Ephesians 4:26).	Sometimes it's just easier to unplug from reality and escape. Music, games, shows, and social media can be an escape and sometimes I self-medicate stress, anger, shame, or boredom with Drool Tech.
T: TIME?	I set time limits and keep them. I make tech wait for me. I use tech during assigned times and then can walk away.	Even when I'm not around tech I think about it. I make time for tech even if it isn't convenient (stay up late, wake up early, sneak a few minutes walking down the hall).

Endnotes

Introduction

1. There is no such thing as… Lindsay Holmes, "'Technology Addiction' Isn't Real, But You Can Have An Obsession," HuffPost, February 21, 2017, https://www.huffpost.com/entry/technology-addiction-isnt-real-but-you-can-have-an-obsession_n_58a5cb4ae4b037d17d25799f.
2. Tech addiction is growing… World Health Organization, "International Classification of Diseases for Mortality and Morbidity Statistics (11th Revision). Chapter 6C51: Gaming Disorder," World Health Organization - Family of International Classifications, accessed 9.3.2023, https://icd.who.int/dev11/f/en#/http%3A%2F%2Fid.who.int%2Ficd%2Fentity%2F1448597234.
3. Tech addiction is growing… "Addictive Behaviours: Gaming Disorder," World Health Organization, October 22, 2020, https://www.who.int/news-room/questions-and-answers/item/addictive-behaviours-gaming-disorder.
4. Tech addiction is growing… Nazia Darvesh et al.,

"Exploring the Prevalence of Gaming Disorder and Internet Gaming Disorder: A Rapid Scoping Review," Systematic Reviews 9, no. 1 (April 2, 2020), https://doi.org/10.1186/s13643-020-01329-2.

5. Tech addiction is growing… Nicholas Kardaras, Glow Kids: How Screen Addiction Is Hijacking Our Kids—and How to Break the Trance (New York: St. Martin's Griffin, 2016), 61–62.
6. Tech addiction is growing… Maria Konnikova, "Is Internet Addiction a Real Thing?," The New Yorker, November 26, 2014, https://www.newyorker.com/science/maria-konnikova/internet-addiction-real-thing.
7. Technology is good for… Lisa Geng, "Video Games May Help With Academics, Motor Skills, Behavior And Much More," Pursuit of Research, May 27, 2022, https://pursuitofresearch.org/2014/02/19/video-games-as-therapy/.
8. Technology is good for… Drew Guarini, "9 Ways Video Games Can Actually Be Good For You," HuffPost, December 7, 2017, https://www.huffpost.com/entry/video-games-good-for-us_n_4164723.
9. Technology is good for… Brian D. Glass, W. Todd Maddox, and Bradley C. Love, "Real-Time Strategy Game Training: Emergence of a Cognitive Flexibility Trait," PLOS ONE 8, no. 8 (August 7, 2013): e70350, https://doi.org/10.1371/journal.pone.0070350.
10. Technology impairs…Jessica S. Mendoza et al., "The Effect of Cellphones on Attention and Learning: The Influences of Time, Distraction, and Nomophobia," Computers in Human Behavior 86

(September 1, 2018): 52–60, https://doi.org/10.1016/j.chb.2018.04.027.

11. Technology impairs… Louis-Philippe Beland and Richard Murphy, "Ill Communication: Technology, Distraction & Student Performance," Labour Economics 41 (August 1, 2016): 61–76, https://doi.org/10.1016/j.labeco.2016.04.004.
12. Technology impairs… Robert Weis and Brittany C. Cerankosky, "Effects of Video-Game Ownership on Young Boys' Academic and Behavioral Functioning," Psychological Science 21, no. 4 (February 18, 2010): 463–70, https://doi.org/10.1177/0956797610362670.
13. Gaming improves…Gideon P. Caplovitz and Sabine Kastner, "Carrot Sticks or Joysticks: Video Games Improve Vision," Nature Neuroscience12, no. 5 (May 1, 2009): 527–28, https://doi.org/10.1038/nn0509-527.
14. Gaming improves… Brian Handwerk, "Video Games Improve Vision, Study Says," Science, May 4, 2009, https://www.nationalgeographic.com/science/article/news-video-games-vision-correction-sight-medicine.
15. Screens ruin… Stephanie Watson, "What Is Computer Vision Syndrome?," WebMD, July 22, 2008, https://www.webmd.com/eye-health/computer-vision-syndrome.
16. Screens ruin… Mark Rosenfield, "Computer Vision Syndrome: A Review of Ocular Causes and Potential Treatments," Ophthalmic and Physiological Optics 31, no. 5 (April 12, 2011): 502–15, https://doi.org/10.1111/j.1475-1313.2011.00834.x.
17. Screens ruin…Jason Howland, "Mayo Clinic Minute:

How Staring at Screens Can Affect Your Vision," Mayo Clinic News Network, August 18, 2023, https://newsnetwork.mayoclinic.org/discussion/mayo-clinic-minute-how-staring-at-screens-can-affect-your-vision/.

18. Tech improves attention… Walter R. Boot et al., "The Effects of Video Game Playing on Attention, Memory, and Executive Control," Acta Psychologica 129, no. 3 (November 1, 2008): 387–98, https://doi.org/10.1016/j.actpsy.2008.09.005.

19. Tech causes attention issues… Adrian F. Ward et al., "Brain Drain: The Mere Presence of One's Own Smartphone Reduces Available Cognitive Capacity," Journal of the Association for Consumer Research 2, no. 2 (April 1, 2017): 140–54, https://doi.org/10.1086/691462.

20. Tech causes attention issues… Kevin McSpadden, "You Now Have a Shorter Attention Span than a Goldfish," Time, May 14, 2015, https://time.com/3858309/attention-spans-goldfish/.

21. We have evolved…Yuval Noah Harari, Homo Deus: A Brief History of Tomorrow, (HarperCollins, 2015), 46.

22. We have evolved…Carl Zimmer, "How Google Is Making Us Smarter," Discover Magazine, October 15, 2019, https://www.discovermagazine.com/mind/how-google-is-making-us-smarter.

23. Modern tech inhibits our ability…Nicholas Carr, "How Smartphones Hijack Our Minds," WSJ, October 6, 2017, https://www.wsj.com/articles/how-smartphones-hijack-our-minds-1507307811.

24. Modern tech inhibits our ability…Cal Newport, Deep

Work: Rules for Focused Success in a Distracted World, (New York: Grand Central Publishing, 2016), 44.

25. Modern tech inhibits our ability…Nicholas G. Carr, The Shallows: How the Internet Is Changing the Way We Think, Read and Remember, (W. W. Norton & Company, 2010).
26. Screen time is no different…Ari Brown, Donald L. Shifrin, and David L. Hill, "Beyond 'Turn It off': How to Advise Families on Media Use," AAP News 36, no. 10 (October 1, 2015): 54, https://doi.org/10.1542/aapnews.20153610-54.
27. Screen time negatively impacts…l Lynne Peeples, "Mental Problems Rise with Kids' Screen Time: Study," U.S., October 11, 2010, https://www.reuters.com/article/health-us-mental/mental-problems-rise-with-kids-screen-time-study-idUKTRE69A0GD20101011.
28. Screen time negatively impacts… Patti Neighmond, "Increased Hours Online Correlate With An Uptick In Teen Depression, Suicidal Thoughts," NPR, November 14, 2017, https://www.npr.org/sections/health-shots/2017/11/14/563767149/increased-hours-online-correlate-with-an-uptick-in-teen-depression-suicidal-thou.
29. Bullying and predators aren't…Christine Elgersma, "The Facts about Online Predators Every Parent Should Know," Common Sense Media, July 25, 2017, https://www.commonsensemedia.org/articles/the-facts-about-online-predators-every-parent-should-know.

30. Eleven percent of children experience unwanted... Sheri Madigan et al., "The Prevalence of Unwanted Online Sexual Exposure and Solicitation among Youth: A Meta-Analysis," Journal of Adolescent Health 63, no. 2 (August 1, 2018): 133–41, https://doi.org/10.1016/j.jadohealth.2018.03.012.

31. "Unfettered access to everything, everywhere, at any age has been a disaster." Jonathan Haidt, The Anxious Generation: How the Great Rewiring of Childhood Is Causing an Epidemic of Mental Illness (New York: Penguin Publishing, 2024), 139.

32. The Gospel says you are more sinful...Timothy Keller, "The Meaning of Marriage: Facing the Complexities of Commitment with the Wisdom of God," (New York: Penguin Books, 2011), 44.

33. Keep tech in its proper place...Andy Crouch, The Tech-Wise Family (Grand Rapids, Michigan:BakerBooks, 2017), 15, 29.

Chapter 1: My Tech Testimony

1. "If you've ever asked...", Pastor Steve Schell, Northwest Church, May 15, 2011.

2. "Our hearts are idol factories...Timothy Keller, Counterfeit Gods: The Empty Promises of Money, Sex, and Power, and the Only Hope That Matters, (Dutton, 2009), xvii.

3. The best games...John Hopson, "Behavioral Game Design," Game Developer, April 27, 2001, https://www.gamedeveloper.com/design/behavioral-game-design.

Chapter 2: What do we mean when we say "Tech"?

1. Trevor Noah (2018), Son of Patricia, Netflix, 2018.
2. Main competition isn't other streaming…Rina Raphael, "Netflix CEO Reed Hastings: Sleep Is Our Competition," Fast Company, November 6, 2017, https://www.fastcompany.com/40491939/netflix-ceo-reed-hastings-sleep-is-our-competition.
3. The merging of reasoned information with entertainment…Neil Postman, Amusing Ourselves to Death: Public Discourse in the Age of Show Business. 2010, Penguin Books, xx.
4. Supersize Me…Spurlock, Morgan. 2004. Super Size Me. United States: Roadside Attractions.
5. Facebook, for example… "How Facebook distributes content", Meta Business Help Center, https://www.facebook.com/business/help/718033381901819?id=208060977200861.

Chapter 3: Designed to Distract

1. Operant Conditioning…J. E. R. Staddon and Yaron Niv, "Operant Conditioning," Scholarpedia 3, no. 9 (January 1, 2008): 2318, https://doi.org/10.4249/scholarpedia.2318.
2. Hebbian Theory of Learning…D. O. Hebb, The Organization of Behavior, MIT Press EBooks, 1988.
3. Pre-scientific superstition…Burrhus Frederick Skinner, (1947). 'Superstition' in the pigeon. Journal of Experimental Psychology, 38, 168-172.as cited in "B.

F. Skinner," Department of Psychology, n.d., https://psychology.fas.harvard.edu/people/b-f-skinner.

4. It is the environment, not the mind…Heitzman, Garlen D., and B. F. Skinner. "Reflections on Behaviorism and Society." Contemporary Sociology 8, no. 5 (1979): 775. https://doi.org/10.2307/2065465.

5. No dividing line between man and brute…John B. Watson, "Psychology as the Behaviorist Views It.," Psychological Review 20, no. 2 (March 1, 1913): 158–77, https://doi.org/10.1037/h0074428.

6. Neuroplasticity refers to the ability…Jeffrey Schwartz and Sharon Begley, The Mind and the Brain: Neuroplasticity and the Power of Mental Force, (New York:HarperCollins, 2002), 15.

7. a study of London cab drivers…Ed Yong, "How Acquiring The Knowledge Changes the Brains of London Cab Drivers," December 8, May 3, 2011, https://www.nationalgeographic.com/science/article/acquiring-the-knowledge-changes-the-brains-of-london-cab-drivers. Accessed 9.3.23

8. Highlights the magnitude of what neuroplasticity can mean for medicine…Norman Doidge, The Brain That Changes Itself (United States: Penguin Books, 2007), 1.

9. The basics of behavioral training…Nir Eyal, Hooked: How to Build Habit-Forming Products (United States: Portfolio/Penguin, 2019), iv.

10. Addiction is any time you give up what you need… Cameron Birk, (CEO of Absolute Ministries), April, 19th 2018. https://www.absoluteministries.org.

Chapter 4: Your kid doesn't need a smartphone

1. Doesn't fully kick in until 25…Mariam Arain et al., "Maturation of the Adolescent Brain," Neuropsychiatric Disease and Treatment, April 1, 2013, 449, https://doi.org/10.2147/ndt.s39776.

2. Heavy phone or social media use may also have a cumulative, enduring, and deleterious effect…Jon Haidt, "Get Phones Out of Schools Now," The Atlantic, June 6, 2023, https://www.theatlantic.com/ideas/archive/2023/06/ban-smartphones-phone-free-schools-social-media/674304/

3. Two to six times more likely to cause harm… Jean M. Twenge et al., "Underestimating Digital Media Harm," Nature Human Behaviour 4, no. 4 (April 17, 2020): 346–48, https://doi.org/10.1038/s41562-020-0839-4.

4. More than two hours on the internet are forty-eight percent more likely to self-harm … Jean M. Twenge et al., "Underestimating Digital Media Harm," Nature Human Behaviour 4, no. 4 (April 17, 2020): 346–48, https://doi.org/10.1038/s41562-020-0839-4.

5. Directed attention is the ability to inhibit distractions and sustain attention…Julie Jargon, "TikTok Brain Explained: Why Some Kids Seem Hooked on Social Video Feeds," The Wall Street Journal, April 2, 2022, https://www.wsj.com/articles/tiktok-brain-explained-why-some-kids-seem-hooked-on-social-video-feeds-11648866192.

6. Videos on the platform average between thirty-two to forty-three seconds…L. Ceci, Average TikTok video length from August 2022 to January 2023, by number

of video views, Statista, accessed 9.5.23, https://www.statista.com/statistics/1372569/tiktok-video-duration-by-number-of-views/

7. TikTok users averaging ninety-five minutes of viewing each day...Rohit Shewale, "46 TikTok Statistics For 2024 (Users, Creators & Revenue)," demandsage, December 26, 2023, https://www.demandsage.com/tiktok-user-statistics/

8. One of the top apps…Emily A. Vogels, Risa Gelles-Watnick and Navid Mssarat, "Teens, Social Media and Technology 2022 | Pew Research Center," Pew Research Center: Internet, Science & Tech, December 15, 2022, https://www.pewresearch.org/internet/2022/08/10/teens-social-media-and-technology-2022/.

9. fifty-eight percent of teens visiting daily…Emily A. Vogels, Risa Gelles-Watnick, Navid Massarat, "Teens, Social Media and Technology 2022 | Pew Research Center." https://www.pewresearch.org/internet/2022/08/10/teens-social-media-and-technology-2022/

10. "real world"…Jonathan Haidt, The Anxious Generation: How the Great Rewiring of Childhood Is Causing an Epidemic of Mental Illness (New York: Penguin Publishing, 2024), 9-10.

Jonathan Haidt's definition of the real world versus the digital world is worth quoting in full: "When I talk about the 'real world,' I am referring to relationships and social interactions characterized by four features that have been typical for millions of years:

- They are embodied, meaning that we use our bodies to communicate, we are conscious of the bodies of others, and we respond to the bodies of others both consciously and unconsciously.
- They are synchronous, which means they are happening at the same time, with subtle cues about timing and turn taking.
- They involve primarily one-to-one or one-to-several communication, with only one interaction happening at a given moment.
- They take place within communities that have a high bar for entry and exit, so people are strongly motivated to invest in relationships and repair rifts when they happen.

In contrast, when I talk about the 'virtual world,' I am referring to relationships and interactions characterized by four features that have been typical for just a few decades:

- They are disembodied, meaning that no body is needed, just language. Partners could be (and already are) artificial intelligences (AIs).
- They are heavily asynchronous, happening via text-based posts and comments. (A video call is different; it is synchronous.)
- They involve a substantial number of one-to-many communications, broadcasting to a potentially vast audience. Multiple interactions can be happening in parallel.
- They take place within communities that have a low bar for entry and exit, so people can block others or just quit when they are not pleased. Communities

tend to be short-lived, and relationships are often disposable."

11. When we send our children into places like Roblox unattended...Sean Murray, "Roblox Is A Cesspool Of Child Abuse According To New Report," TheGamer, December 13, 2021, https://www.thegamer.com/roblox-child-abuse-people-make-games-report/.
12. The more time young children spend online...Emily Vogels, "Teens and Cyberbullying 2022 | Pew Research Center," Pew Research Center: Internet, Science & Tech, December 15, 2022, https://www.pewresearch.org/internet/2022/12/15/teens-and-cyberbullying-2022/.
13. Making them porn resilient..."Fight Porn — Blog — Greta Eskridge," Greta Eskridge, November 17, 2021, http://www.gretaeskridge.com/blog/tag/Fight+Porn.
14. My favorite resources here...Kristen A. Jensen, Good Pictures, Bad Pictures, (Glen Cove Press, 2014).

Chapter 5: Social Media

1. Teenagers spend 8.5 hours a day...Victoria Rideout, Alanna Peebles, Supreet Mann, and Michael Robb, "Common Sense Census: Media Use by Tweens and Teens, 2021," commonsensemedia.org, accessed September 5, 2023, https://www.commonsensemedia.org/sites/default/files/research/report/8-18-census-integrated-report-final-web_0.pdf.
2. Social media platforms are often designed...Vivek Murthy, 2023, "Social Media and Youth Mental Health: The US Surgeon General's Advisory." Surgeongeneral.Gov, 2023., pg.9. Accessed September 6, 2023. https://

www.hhs.gov/sites/default/files/sg-youth-mental-health-social-media-advisory.pdf.

3. It's actually not surprising…Jonathan Haidt, The Anxious Generation: How the Great Rewiring of Childhood Is Causing an Epidemic of Mental Illness (New York: Penguin Publishing, 2024), 91.

4. Race to the bottom of the Brainstem…"The Social Dilemma." 2020. Jeff Orlowski, Larissa Rhodes. Accessed September 6, 2023. https://www.thesocialdilemma.com/.

5. 95% of US teens use social media…Pew Research Center. 2023. "Teens and Social Media: Key Findings from Pew Research Center Surveys | Pew Research Center." June 12, 2023. https://www.pewresearch.org/short-reads/2023/04/24/teens-and-social-media-key-findings-from-pew-research-center-surveys/

6. A host of tricks including…Griffiths, Mark. 2018. "Adolescent Social Networking: How Do Social Media Operators Facilitate Habitual Use?" Education and Health 36 (3): 66–69. https://sheu.org.uk/sheux/EH/eh363mdg.pdf

7. 1/3 of social media…Hunt Allcott, Matthew Gentzkow, & Lena Song, Digital Addiction American Economic Review, 112, no. 7, (2022): 2424–63. https://doi.org/10.1257/aer.20210867 as cited in Murthy, Vivek. 2023. "Social Media and Youth Mental Health: The US Surgeon General's Advisory." Surgeongeneral.Gov. US Department of Health and Human Services. pg.9–10. Accessed September 6, 2023.

8. 1/3 of Jacqueline Nesi, Supreet Mann, and Michael

Robb, "Teens and mental health: How girls really feel about social media.", commonsensemedia.org, accessed September 5, 2023, https://www.commonsensemedia.org/sites/default/files/research/report/how-girls-really-feel-about-social-media-researchreport_web_final_2.pdf, as cited in Murthy, Vivek. 2023. "Social Media and Youth Mental Health: The US Surgeon General's Advisory." US Department of Health and Human Services. pg.9–10. Accessed September 6, 2023.

9. 3.5 hours a day…Luca Braghieri, Ro'ee Levy, & Alexi Makarin, "Social Media and Mental Health," American Economic Review, 112, no.11 (2022):3660-3693. https://pubs.aeaweb.org/doi/abs/10.1257/aer.20211218 as cited in Murthy, Vivek. 2023. "Social Media and Youth Mental Health: The US Surgeon General's Advisory." US Department of Health and Human Services. pg.9-10. Accessed September 6, 2023.

10. Use social media almost constantly…Emily A. Vogels, Risa Gelles-Watnick, Navid Massarat, "Teens, Social Media and Technology 2022 | Pew Research Center." https://www.pewresearch.org/internet/2022/08/10/teens-social-media-and-technology-2022/ as cited in Murthy, Vivek. 2023. "Social Media and Youth Mental Health: The US Surgeon General's Advisory." Surgeongeneral.Gov. US Department of Health and Human Services. pg.9–10. Accessed September 6, 2023.

11. Stewarding our lives unto the glory of God…Ephesians 4:22–24, 5:15–17, 2 Corinthians 5:17, Mark 12:30.

12. In 2021 nearly three out of five teenage girls…"U.S. Teen Girls Experiencing Increased Sadness and

Violence." 2023. Centers for Disease Control and Prevention. February 13, 2023, 2. https://www.cdc.gov/media/releases/2023/p0213-yrbs.html.

13. One third of teen boys…"U.S. Teen Girls Experiencing Increased Sadness and Violence." 2023. Centers for Disease Control and Prevention. February 13, 2023. 2 https://www.cdc.gov/media/releases/2023/p0213-yrbs.html.
14. iGen…Jean Twenge, iGen: Why Today's Super-Connected Kids are Growing Up Less Rebellious, More Tolerant, Less Happy–and Completely Unprepared for Adulthood—and What That Means for the Rest of Us. (Atria Books, 2017), 2.
15. Social media is a substantial cause…Jon Haidt, "Social Media Is a Major Cause of the Mental Illness Epidemic in Teen Girls. Here's The Evidence.," After Babel, February 22, 2023, accessed September 6, 2023, https://jonathanhaidt.substack.com/p/social-media-mental-illness-epidemic.
16. Personally, based on the data I've seen…Vivek Murthy, Surgeon General says 13 is 'too early' to join social media. https://www.cnn.com/2023/01/29/health/surgeon-general-social-media/index.html
17. Introduction of social media to youth…Fuhrmann, D., Knoll, L. J., & Blakemore, S. J. (2015). Adolescence as a Sensitive Period of Brain Development. Trends in cognitive sciences, 19(10), 558–566. https://doi.org/10.1016/j.tics.2015.07.008 as cited in Vivek Murthy. 2023. "Social Media and Youth Mental Health: The US Surgeon General's Advisory." Surgeongeneral.Gov, 2023., pg.7. Accessed September 6, 2023.

18. Introduction of social media to youth…Blakemore, S. J., & Mills, K. L. (2014). Is adolescence a sensitive period for sociocultural processing?. Annual review of psychology, 65, 187–207. https://doi.org/10.1146/annurev-psych-010213-115202

19. Introduction of social media to youth…Romer D. (2010). Adolescent risk taking, impulsivity, and brain development: implications for prevention. Developmental psychobiology, 52(3), 263–276. https://doi.org/10.1002/dev.20442

20. Introduction of social media to youth…National Academies of Sciences, Engineering, and Medicine (NASEM). (2019). The Promise of Adolescence: Realizing Opportunity for All Youth. Washington, DC: The National Academies Press. https://doi.org/10.17226/25388

21. Introduction of social media to youth…Kessler, R. C., Amminger, G. P., Aguilar-Gaxiola, S., Alonso, J., Lee, S., & Ustün, T. B. (2007). Age of onset of mental disorders: a review of recent literature. Current opinion in psychiatry, 20(4), 359–364. https://doi.org/10.1097/YCO.0b013e32816ebc8c15

22. The roll-out of the platform was associated…Braghieri, L., Levy, R., & Makarin, A. (2022). Social Media and Mental Health. American Economic Review, 112(11), 3660–3693. https://pubs.aeaweb.org/doi/abs/10.1257/aer.20211218 as cited in Vivek Murthy. 2023. "Social Media and Youth Mental Health: The US Surgeon General's Advisory." Surgeongeneral.Gov, 2023., pg.7. Accessed September 6, 2023.

23. There is one giant, obvious, international, and gendered

cause…Jon Haidt, "Social Media Is a Major Cause of the Mental Illness Epidemic in Teen Girls. Here's The Evidence.," After Babel, February 22, 2023, https://www.afterbabel.com/p/social-media-mental-illness-epidemic

24. FOMO is associated with…Fioravanti, G., Casale, S., Benucci, S. B., Prostamo, A., Falone, A., Ricca, V., & Rotella, F. (2021). Fear of missing out and social networking sites use and abuse: A meta-analysis. Computers in Human Behavior. As cited in Vivek Murthy. 2023. "Social Media and Youth Mental Health: The US Surgeon General's Advisory." Surgeongeneral.Gov, 2023., pg.10. Accessed September 6, 2023.

25. Thirty-two percent of teen girls…Georgia Wells, Jeff Horwitz, Deepa Seetharaman. Facebook Knows Instagram Is Toxic for Teen Girls, Company Documents Show. (Wall Street Journal, September 14, 2021). Accessed 9.6.23. https://www.wsj.com/articles/facebook-knows-instagram-is-toxic-for-teen-girls-company-documents-show-11631620739

26. 40% of Instagram's user base…Georgia Wells, Jeff Horwitz, Deepa Seetharaman. Facebook Knows Instagram Is Toxic for Teen Girls, Company Documents Show. (Wall Street Journal, September 14, 2021). Accessed 9.6.23. https://www.wsj.com/articles/facebook-knows-instagram-is-toxic-for-teen-girls-company-documents-show-11631620739

27. Twenty-five percent are spending five plus…Braghieri, L., Levy, R., & Makarin, A. (2022). Social Media and Mental Health. American Economic Review, 112(11), 3660-3693.

28. A mere two hours a day…Jean M. Twenge et al., “Increases in Depressive Symptoms, Suicide-Related Outcomes, and Suicide Rates among U.S. Adolescents after 2010 and Links to Increased New Media Screen Time,” Clinical Psychological Science 6, no. 1 (November 14, 2017): 3–17, https://doi.org/10.1177/2167702617723376.

29. Like Christian in Pilgrim’s Progress… Bunyan, “The Pilgrim’s Progress,” in Oxford University Press EBooks, 1678, https://doi.org/10.1093/oseo/instance.00017888.

30. Remember the words of the Surgeon General…Vivek Murthy, Surgeon General says 13 is ‘too early’ to join social media. https://www.cnn.com/2023/01/29/health/surgeon-general-social-media/index.html.

Chapter 6: Video Games

1. The average gamer is thirty-two…Gaming Stats, 2022, ESA, https://www.theesa.com/resource/2022-essential-facts-about-the-video-game-industry/

2. More gamers over the age of 45 than there are under the age of 18…Gaming Stats (2022), ESA, https://www.theesa.com/resource/2022-essential-facts-about-the-video-game-industry/

3. As an example, an ESRB.org search for Mario Kart 8 Deluxe…Mario Kart 8 Deluxe (2024), ESRB, https://www.esrb.org/ratings/34765/mario-kart-8-deluxe/

4. Archeologists have found board games…“Oldest Known Gaming Tokens Dug up in Bronze Age Turkish Graves,” NBC News, August 14, 2013, https://www.

nbcnews.com/sciencemain/oldest-known-gaming-tokens-dug-bronze-age-turkish-graves-6c10920354.

5. 1.175 million players…Matija Ferjan, "Latest World of Warcraft Player Count & Subscription Numbers (2023)," HeadphonesAddict, August 15, 2023, https://headphonesaddict.com/world-of-warcraft-player-count/#World-of-Warcraft-player-count. Accessed 9.7.23

6. 2 hours to beat…"How Long Is Super Mario Bros.? | HowLongToBeat," n.d., https://howlongtobeat.com/game/9371.

Chapter 7: Pornography

1. Porn does more than just amuse; it trains…Fight the New Drug, "How Porn Can Change the Brain," Fight the New Drug, March 21, 2023, https://fightthenewdrug.org/how-porn-can-change-the-brain/.

2. Porn does more than just amuse; it trains…Fight the New Drug, "How Porn Can Become an Escalating Behavior," Fight the New Drug, March 21, 2023, https://fightthenewdrug.org/how-porn-can-become-an-escalating-behavior/.

3. Porn manipulates the way our brain works…Fight the New Drug, "How Porn Can Contribute to an Unhealthy Cycle of Stress," Fight the New Drug, October 3, 2022, https://fightthenewdrug.org/how-porn-can-contribute-to-an-unhealthy-cycle-of-stress/.

4. Porn does more than just amuse; it trains…Fight the New Drug, "How Porn Can Affect the Brain like a Drug," Fight the New Drug, March 21, 2023, https://

fightthenewdrug.org/how-porn-can-affect-the-brain-like-a-drug/.

5. Pornography can take our brains' natural stimuli… Voon, V., et al. (2014). Neural Correlates of Sexual Cue Reactivity in Individuals with and without Compulsive Sexual Behaviors, PLoS ONE, 9(7), e102419. doi:10.1371/journal.pone.0102419 as cited in Fight the New Drug, how Porn Can Change the Brain. https://fightthenewdrug.org/how-porn-can-change-the-brain/ accessed 9.7.23

6. Helicopter parenting can actually make matters worse…Isabel Hogben, "I Had a Helicopter Mom. I Found Pornhub Anyway.," Https://Www.Thefp.Com/, August 29, 2023, https://www.thefp.com/p/why-are-our-fourth-graders-on-pornhub.

7. Somewhere between the ages of 8 and 12…Sam Black, "190. How To Talk With Our Children about Porn." gospeltechpodcast.com, September 19, 2023.

8. Even the Amish struggle…Justine Sharrock, "Why Amish Teens Love Facebook," BuzzFeed News, May 28, 2013, https://www.buzzfeednews.com/article/justinesharrock/why-amish-teens-love-facebook.

9. Our responsibility to them is not…Madeleine L'Engle, A Circle of Quiet, 1971, https://www.amazon.com/Circle-Quiet-Madeleine-LEngle/dp/0062545035.

10. Pornography changes the brain…develops unhealthy appetites…Deirdre Barret. (2010). Supernormal stimulus. New York: W. W. Norton & Company. Fight the New Drug, "How Porn Can Change the Brain," Fight the New Drug, March 21, 2023. https://

fightthenewdrug.org/how-porn-can-change-the-brain/.

11. Creates a culture of abuse…Tré Goins-Phillips, "'As a Woman, I Think Porn Is a Disgrace': Billie Eilish Says She Started Watching Porn at 11, Admits It 'Destroyed' Her Brain | CBN," CBN, December 15, 2021, https://www2.cbn.com/news/news/woman-i-think-porn-disgrace-billie-eilish-says-she-started-watching-porn-11-admits-it.
12. Living and active… Hebrews 4:12 "For the word of God is living and active, sharper than any two-edged sword, piercing to the division of soul and of spirit, of joints and of marrow, and discerning the thoughts and intentions of the heart."
13. Raising them up…Proverbs 22:6 "Train up a child in the way he should go: And when he is old, he will not depart from it.
14. Knows the hairs on our heads…Luke 12:7, "And the very hairs on your head are all numbered. So don't be afraid; you are more valuable to God than a whole flock of sparrows."
15. The thoughts inside them…1 Chronicles 28:9, "And you, Solomon my son, know the God of your father and serve him with a whole heart and with a willing mind, for the LORD searches all hearts and understands every plan and thought. If you seek him, he will be found by you, but if you forsake him, he will cast you off forever."
16. Thoughts matter as much as our actions…Matthew 5:22, " But I say unto you, That whosoever is angry with his brother without a cause shall be in danger of

the judgment: and whosoever shall say to his brother, Raca, shall be in danger of the council: but whosoever shall say, Thou fool, shall be in danger of hell fire."

17. Big enough to forgive…Matthew 11:28, "28 "Come to me, all you who are weary and burdened, and I will give you rest. 29 Take my yoke upon you and learn from me, for I am gentle and humble in heart, and you will find rest for your souls. 30 For my yoke is easy and my burden is light."

18. Big enough to forgive…John 3:15–17, "For God so loved the world, that he gave his only begotten Son, that whosoever believeth in him should not perish, but have everlasting life. For God sent not his Son into the world to condemn the world; but that the world through him might be saved."

19. The one who has overcome…John 16:33, "I have said these things to you, that in me you may have peace. In the world you will have tribulation. But take heart; I have overcome the world."

20. Called them to good works…Ephesians 2:10, For we are God's handiwork, created in Christ Jesus to do good works, which God prepared in advance for us to do."

21. A warm and communicative parent-child relationship… Patricia M. Greenfield, "Inadvertent Exposure to Pornography on the Internet: Implications of Peer-to-Peer File-Sharing Networks for Child Development and Families," Journal of Applied Developmental Psychology 25, no. 6 (November 1, 2004): 741–50, https://doi.org/10.1016/j.appdev.2004.09.009.

22. Bath time is a great opportunity to have those drip, drip, drip…Megan Flo, Birds & Bees (@Birds__Bees). "Bathtime is a great opportunity to have those drip, drip, drip conversations with your little kids." Instagram, April 17, 2023. https://www.instagram.com/p/CrKIrd8ME2B/
23. People not wearing clothes being unkind with their bodies …Kristen A. Jensen, Good Pictures, Bad Pictures, (Glen Cove Press, 2014)
24. Instead of hiding it…Chris McKenna, "117. Let's Talk Safe Tech, A Conversation with Chris McKenna of Protect Young Eyes." gospeltechpodcast.com, May 17, 2022. https://podcasts.apple.com/us/podcast/gospel-tech/id1498508882?i=1000561743928.

Chapter 8: Assessing Healthy Tech Use

1. What you give your attention…John Mark Comer, The Ruthless Elimination of Hurry (United States: WaterBrook, 2019), 54.
2. The Blue Light emitted from screens…Marcia Ines Silvani, Robert Werder, and Catherine Perret, "The Influence of Blue Light on Sleep, Performance and Wellbeing in Young Adults: A Systematic Review," Frontiers in Physiology 13 (August 16, 2022), https://doi.org/10.3389/fphys.2022.943108.
3. Using a smartphone in bed…Saba Kheirinejad et al., "'Leave Your Smartphone out of Bed': Quantitative Analysis of Smartphone Use Effect on Sleep Quality," Personal and Ubiquitous Computing 27, no. 2 (November 9, 2022): 447–66, https://doi.org/10.1007/

s00779-022-01694-w.

4. Notifications are going off throughout the night…Karla Klein Murdock et al., "Nighttime Notifications and Compulsivity Illuminate the Link between Emerging Adults' Cellphone Use and Sleep-Related Problems.," Psychology of Popular Media Culture 8, no. 1 (January 1, 2019): 12–21, https://doi.org/10.1037/ppm0000156.
5. The fear of missing out paired with a false sense of urgency…Karla Klein Murdock, "Texting While Stressed: Implications for Students' Burnout, Sleep, and Well-Being.," Psychology of Popular Media Culture 2, no. 4 (October 1, 2013): 207–21, https://doi.org/10.1037/ppm0000012.
6. The fear of missing out paired with a false sense of urgency…Sue K. Adams et al., "The Young and the Restless: Socializing Trumps Sleep, Fear of Missing out, and Technological Distractions in First-Year College Students," International Journal of Adolescence and Youth 22, no. 3 (May 11, 2016): 337–48, https://doi.org/10.1080/02673843.2016.1181557.
7. The fear of missing out paired with a false sense of urgency…Ursula Oberst et al., "Negative Consequences from Heavy Social Networking in Adolescents: The Mediating Role of Fear of Missing Out," Journal of Adolescence 55, no. 1 (December 27, 2016): 51–60, https://doi.org/10.1016/j.adolescence.2016.12.008.
8. Screen Time assessment…Chris McKenna, "IOS 16 Makes Big Changes to Screen Time Parental Controls," Protect Young Eyes, September 12, 2022, https://

protectyoungeyes.com/ios-16-makes-big-changes-screen-time-parental-controls/.

9. Putting your phone out of your bedroom increases your happiness…Niall Hughes and Jolanta Burke, "Sleeping with the Frenemy: How Restricting 'Bedroom Use' of Smartphones Impacts Happiness and Wellbeing," Computers in Human Behavior 85 (August 1, 2018): 236–44, https://doi.org/10.1016/j.chb.2018.03.047.

10. Young adult needs nine hours…Max Hirshkowitz et al., "National Sleep Foundation's Updated Sleep Duration Recommendations: Final Report," Sleep Health 1, no. 4 (December 1, 2015): 233–43, https://doi.org/10.1016/j.sleh.2015.10.004.

11. When sleep is abundant…Matthew Walker, Why We Sleep: Unlocking the Power of Sleep and Dreams (New York: Simon & Schuster, 2017), 316.

12. A lack of sleep ahs been linked to an increase risk of depression and suicide…Matthew Walker, Why We Sleep: Unlocking the Power of Sleep and Dreams (New York: Simon & Schuster, 2017), 133, 148, 151.

13. Too little causes our bodies to go into flight-or-flight mode…Matthew Walker, Why We Sleep: Unlocking the Power of Sleep and Dreams (New York: Simon & Schuster, 2017), 168.

14. The general rule is less than two hours a day on screens…Jean M. Twenge et al., "Increases in Depressive Symptoms, Suicide-Related Outcomes, and Suicide Rates among U.S. Adolescents after 2010 and Links to Increased New Media Screen Time," Clinical Psychological Science 6, no. 1 (November 14, 2017):

3–17, https://doi.org/10.1177/2167702617723376.

It's worth quoting the research in full for the record: "Electronic device use showed a clear exposure response relationship with having at least one suicide risk factor, in both bivariate analyses and those including demographic controls (see Fig. 2). Risks became elevated after 2 hours or more a day of electronic device use, when 33% of adolescents had at least one suicide related outcome."

Chapter 9: Making Tech Safe at Home

1. Keep tech in its proper place…Andy Crouch, The Tech-Wise Family (Grand Rapids, Michigan:BakerBooks, 2017), 15.

Chapter 10: Adopting New Tech

1. Our thoughts, brain, and even genes adapt to our choices…Norman Doidge, The Brain That Changes Itself (United States: Penguin Books, 2007),.
2. Our thoughts, brain, and even genes adapt to our choices…Jorge Alejandro Alegría-Torres, Andrea Baccarelli, and Valentina Bollati, "Epigenetics and Lifestyle," Epigenomics 3, no. 3 (June 1, 2011): 267–77, https://doi.org/10.2217/epi.11.22.
3. Our thoughts, brain, and even genes adapt to our choices…Samir Zakhari, "Alcohol Metabolism and Epigenetics Changes," PubMed Central (PMC), 2013, https://www.ncbi.nlm.nih.gov/pmc/articles/PMC3860421/.

Conclusion: Raising Healthy Youth in a Tech World

1. Raise your child…"Train up a child in the way he should go; even when he is old he will not depart from it." (Pr 22:6)
2. In the fear and admonition… "Fathers, do not provoke your children to anger, but bring them up in the discipline and instruction of the Lord." (Ephesians 6:4)
3. "Lord, make me good…St. Augustine, "The Confessions of St. Augustine," in The Project Gutenberg eBook, 2023, section 8.10, https://www.gutenberg.org/cache/epub/3296/pg3296-images.html.
4. That everything we set our minds on is…"Finally, brethren, whatever is true, whatever is honorable, whatever is right, whatever is pure, whatever is lovely, whatever is of good repute, if there is any excellence and if anything worthy of praise, dwell on these things." (Phil 4:8)
5. We were first loved…"We love because he first loved us." (1 Jn 4:19).

Additional Resources

Check out **Gospel Tech**, the weekly podcast for parents who are feeling outpaced and overwhelmed as they raise children in a tech-saturated world.

Our goal: To equip parents with the tools, resources, and confidence they need to raise kids who love God and use tech (and not the other way around).

Finally, visit GospelTech.net to learn more about our ministry and be sure to follow **@lovegodusetech** on Facebook and Instagram for regular reminders and encouragement.

The Gospel Tech Podcast

Speaking Inquiries for Churches, Schools, and Events

To book Nathan to speak at your event,
please visit **GospelTech.net/Speaking**

Also from Lion Press

In *The Graceful Disconnect: 25 Days to a Quieter Heart in a Tech-Saturated World*, Anna Sutherland offers a refreshing guide for parents seeking to cultivate a deeper connection with their families and faith.

The Graceful Disconnect is a blend of personal anecdotes, biblical insights, and practical challenges designed to help readers take a step back from technology and reconnect with what truly matters. Through a series of daily reflections and actionable tech challenges, readers are encouraged to create a home environment where tech is a tool, not a tyrant, and where love, adventure, and genuine relationships can be reclaimed and redeemed for the glory of God.

Visit LionPress.org to learn more.